PEAS AND BEANS

CROP PRODUCTION SCIENCE IN HORTICULTURE SERIES

Series Editor: Jeff Atherton, Professor of Tropical Horticulture, University of the West Indies, Barbados

This series examines economically important horticultural crops selected from the major production systems in temperate, subtropical and tropical climatic areas. Systems represented range from open field and plantation sites to protected plastic and glass houses, growing rooms and laboratories. Emphasis is placed on the scientific principles underlying crop production practices rather than on providing empirical recipes for uncritical acceptance. Scientific understanding provides the key to both reasoned choice of practice and the solution of future problems.

Students and staff at universities and colleges throughout the world involved in courses in horticulture, as well as in agriculture, plant science, food science and applied biology at degree, diploma or certificate level, will welcome this series as a succinct and readable source of information. The books will also be invaluable to progressive growers, advisers and end-product users requiring an authoritative, but brief, scientific introduction to particular crops or systems. Keen gardeners wishing to understand the scientific basis of recommended practices will also find the series very useful.

The authors are all internationally renowned experts with extensive experience of their subjects. Each volume follows a common format covering all aspects of production, from background physiology and breeding, to propagation and planting, through husbandry and crop protection, to harvesting, handling and storage. Selective references are included to direct the reader to further information on specific topics.

PEAS AND BEANS

Anthony J. Biddle

*Former Technical Director of Processors and
Growers Research Organisation
Peterborough
UK*

CABI is a trading name of CAB International

CABI	CABI
Nosworthy Way	745 Atlantic Avenue
Wallingford	8th Floor
Oxfordshire OX10 8DE	Boston, MA 02111
UK	USA

Tel: +44 (0)1491 832111
Fax: +44 (0)1491 833508
E-mail: info@cabi.org
Website: www.cabi.org

T: +1 (617)682-9015
E-mail: cabi-nao@cabi.org

A catalogue record for this book is available from the British Library, London, UK.

Library of Congress Cataloging-in-Publication Data

Names: Biddle, A. J., author.
Title: Peas and beans / Anthony Biddle.
Other titles: Crop production science in horticulture ; 25.
Description: Boston, MA : CABI, [2017] | Series: Crop production science in horticulture series ; 25
Identifiers: LCCN 2017007118 (print) | LCCN 2017009156 (ebook) | ISBN 9781780640914 (pbk.) | ISBN 9781780640921 | ISBN 9781786393098
Subjects: LCSH: Peas. | Beans.
Classification: LCC SB343 .B53 2017 (print) | LCC SB343 (ebook) | DDC 633.3/7--dc23
LC record available at https://lccn.loc.gov/2017007118

ISBN-13: 978 1 78064 091 4

Commissioning editor: Rachael Russell
Editorial assistant: Alexandra Lainsbury
Production editor: Tim Kapp

Typeset by AMA DataSet Ltd, Preston, UK
Printed and bound in the UK by CPI Group (UK) Ltd, Croydon, CR0 4YY

CONTENTS

PREFACE

Peas and beans represent just a small section of the large family of Fabaceae that contains many hundreds of species, from tall trees to soil-clinging shrubby plants. They have been successful in colonizing many areas of the planet due to their adaptability to grow in a range of conditions from dry hostile semi-deserts to tropical rain forests. However, the main feature that humans have exploited in domesticating those species that could be of most use is their successful symbiotic relationship with soil-borne *Rhizobium* bacteria that allow the plants to produce their own supply of nitrogen nutrition, thereby extending the range of growing habitats to those soils that are inherently low in natural nutrients.

Of this huge group of plants, varieties of peas and beans have been developed from wild ancestors to provide grain crops that are high in food value. Although not understood by the early cultivators of these crops, the seeds contain some of the highest levels of proteins of all crop plants. It is no wonder that humanity has found that their production has provided a rich source of food that is easily stored once dried and can be used in a wide variety of foodstuffs for humans and their animals.

This book has been written to include the main types of peas and beans now grown on a significant scale and mostly commercially produced in large-area farming systems. Whilst there are several types and species of *Pisum*, *Vicia* and *Phaseolus* worldwide, the chapters herein purposely concentrate on peas as *Pisum sativum* and beans as *Vicia faba* and *Phaseolus vulgaris* and confine the species to those grown in cool temperate regions. Although there is some reference to small-scale subsistence farming of these crops, the contents are mainly restricted to the types of crop and their production on a large scale.

I have based much of the information in this book on my own 40-plus years of working in applied research, specializing in peas and beans and in particular as a plant pathologist and agronomist with the Processors and Growers Research Organisation based near Peterborough in Cambridgeshire. During this time I enjoyed close contact with the scientific community and also commercial growers, food processors, merchants and plant breeders in the UK, Europe and the USA, as well as visiting producers in all of these countries and the Middle East and Australia. The chapters therefore deal with the crops in detail as to the agronomy, crop protection and their use in processed food for humans and livestock, but chapters on breeding and physiology have been assisted greatly by colleagues in other research institutes and universities. There are a number of scientific books written by specialists detailing aspects of physiology, genetics and breeding in depth, and although I have included references to the main areas of development in each section, the chapters have been very much summarized for the purposes of this book.

I have aimed this book towards an audience of practical agronomists, students and others involved in the production of peas and beans, concentrating on aspects of husbandry and means of controlling pests, diseases and disorders. In addition, there are details of the characteristics of the varieties of peas and beans desired by producers to maximize their potential for variety improvement by breeders and to show that there is a big future for these crops in supplying food and refined food fractions of high quality and value for humans and their livestock and in the relatively new art of aquaculture. As the world population increases and developing countries become more demanding for meat and dairy-based food, the importance of peas and beans becomes greater both as an additional vegetable food source and as a source of nutritiously healthy products and reducing the reliance on animal proteins for food or for livestock feed. The continuing demand for vegetable protein that is more sustainable than conventional cereals or oilseed crops and that reduces the risks associated with nitrate leaching and nitrous oxide emission is a key factor in promoting peas and beans in temperate cool-season climates and growing areas. It is hoped that this book will provide a useful background to all engaged in production or research of these valuable legume crops.

The book has been a very long time in preparation but I would not have been able to complete the task without the generous help of colleagues and friends. I am particularly indebted to colleagues at PGRO, namely, Steve Belcher, Becky Howard (née Ward) and Jim Scrimshaw, with whom I worked for many years on a whole range of aspects of research into the production of the crops, and the information that I have been able to include in many parts of the book based on our research and practical experiences. Details included in the chapters covering pea breeding and genetics have been greatly helped by Claire Domoney and Mike Ambrose of the John Innes Centre; and Donal O'Sullivan of the University of Reading has helped with critically reviewing the section on faba bean breeding. A lifetime of working with *Phaseolus vulgaris* meant that Colin Leakey was the ideal prime contributor to that section along with relevant parts of other chapters. I am grateful to Pietro Iannetta of James Hutton Institute and Claire Domoney, who have provided detail and expanded the section on nitrogen fixation. On husbandry and harvesting, I am grateful to

Matthew Hayward of Swaythorpe Growers and to Peter Waldock on aspects of fresh market production as well as Keith Costello, formerly of Princes Foods, for aspects of canning and Steve Marx for his description of the freezing operation.

Last but not least, I thank my editor, Rachael Russell, for her patience in waiting for the final manuscript and my wife Linda for prodding me into getting on with the job.

Anthony J. Biddle
Stamford, UK
2017

1

INTRODUCTION TO PEAS AND BEANS

Amongst the world's most important non-cereal food crops, peas and beans are probably the most versatile. They provide a source of protein, are easily stored for long periods and can be consumed as processed or whole food by both humans and livestock. Commonly known as pulse crops or grain legumes, they are widely grown in temperate, subtropical and arid climates all over the world. They can be consumed as fresh vegetables or frozen, canned or dehydrated and also can be harvested as dry seed or pulses, which can be milled for use as a flour, or rehydrated and cooked whole. It seems likely that the adoption of legumes as agricultural crops in part reflects the nutritional balance between legumes and cereal seeds as well as the ability of legumes to break cereal rotations. Because of their ability to fix atmospheric nitrogen through their symbiotic relationship with soil-borne bacteria providing them with sufficient nitrogen for growth, the residue enriches the soil nitrogen supply for the following crop. The diversity of locations where peas and beans have been developed in agriculture is reflected in the diversity of species and varieties currently grown. They are found in agricultural systems throughout the world and have been domesticated in South and Central America, the Middle East, China, India and Africa. More recently they have been introduced to Europe and North America and to other temperate regions in Asia and elsewhere.

PEA AND BEAN FAMILIES

The family Fabaceae has been divided into three main subfamilies: Caesalpinioideae, Mimosoideae and Papilionoideae, but recently Caesalpinioideae has been further subdivided into several lineages, including the tribe Cercideae (Group, 2013) which contains a small group of tropical and temperate woody plants with flowers similar to those of Papilionoideae. The subfamily Mimosoideae includes 82 genera and more than 3200 species. Like Caesalpinioideae, Mimosoideae legumes are mainly tropical woody plants but a few temperate species exist. The woody perennial legumes are widespread and many species within this group have the ability to fix atmospheric nitrogen and as such have colonized many areas of the world where soil nutrients are low. They have also been commercially exploited for their timber or for various extracts such as gums and dyes.

The subfamily Papilionoideae (or Faboideae) is the largest group of legumes, consisting of about 650 genera and nearly 19,000 species (Lewis *et al.*, 2005).

As members of the group of large-seeded legumes in the family Fabaceae, peas and beans fall into several distinct taxonomic groups (Fig. 1.1). Among these, three lineages have a number of groups that contain cultivated crops. The most important group economically is Phaseoleae, which includes soybeans (*Glycine max*), *Phaseolus* species and cowpea (*Vigna unguiculata*). The second group, Genisteae, contains *Lupinus*, of which there are three economi- cally important grain legume species: *Lupinus albus* (white lupin), *Lupinus angustifolium* (narrow-leaved lupin) and *Lupinus luteus* (yellow lupin). The third taxonomic group, Fabeae, contains the so-called cool-season legumes, which include peas (*Pisum sativum*) and faba beans (*Vicia faba*) as well as lentils (*Lens* spp.) and sweet pea (*Lathyrus* spp.).

This book describes the most commonly grown large-seeded legumes used as food crops throughout the world. The species considered are: *Pisum sativum*, in the forms of dried peas (combining peas) and as fresh vegetables (vining peas or fresh market peas); *Vicia faba*, in the dry form as faba beans (or field beans) and the fresh form as broad beans; and *Phaseolus vulgaris*, in the dry form as dried beans and fresh as dwarf French or green (snap) beans.

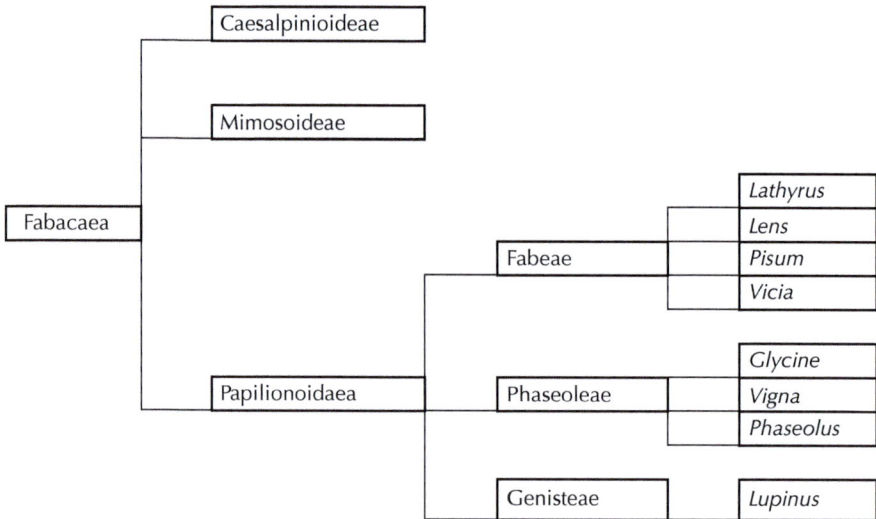

Fig. 1.1. Classification of Fabaceae.

PEAS

History

Pisum sativum is the most commonly cultivated species in temperate climates. It is an annual with seeds that can be harvested in the immature state either as whole pods and consumed as mangetout, or as shelled immature peas that can be consumed fresh, quick frozen, canned or bottled. Peas may also be harvested in the dry mature condition and the dried produce stored for later use, or for seed.

In cultivated use, peas range from an indeterminate growth type with long haulm that can be used as a forage crop, or ensiled for animal feed, to a short-haulmed type with an evenness of maturity of the pods for harvesting on a single occasion, as a vegetable for freezing, canning or for the fresh vegetable market.

Peas probably had their centre of origin in Middle Asia and the central plateau of Ethiopia. By the Bronze Age (c. 3000 BC) they were used by the inhabitants of central Europe and primitive seeds have been found in areas inhabited by Swiss lake dwellers and in caves in central Hungary. Peas were known by the Greeks and Romans and these early types were first mentioned in England after the Norman Conquest (AD 1066).

Fresh peas were popular in the 19th century, when improved varieties were developed by English plant breeders, and in some parts of the world such varieties are called English peas.

Dry-harvested peas were also grown on a wide scale in the 19th century in both Europe and the USA. Dry peas are also grown extensively in more small-scale subsistence systems but the most significant variety development came about by the introduction of mechanical harvesting equipment, firstly (for dry-harvest peas) threshers and latterly the combine harvester and for fresh peas the complete pea harvester known as the viner.

Production

Wet growing conditions can increase the length of haulm and indeterminacy. Botanical determinacy can be achieved by breeding by selection for apical determinacy whereby the flowers and subsequent pods are grouped closely together. Such a type may be more suitable for a single mechanical harvesting operation. There has recently been a further development with modern pea varieties in the production of a stiffer stem, which holds the plant in a more erect position, and '*afila*' types where the leaflets have been converted to tendrils. The tendrils of *afila* types tend to twine together and offer mutual support.

Such morphological changes have greatly improved the agronomic characteristics of peas to enable them to be grown in a wide range of geographical areas and have been exploited by large-scale harvesting operations.

Older less determinate types lend themselves to small-scale and garden production where harvesting is carried out by hand and multiple harvests of the same set of plants can be made to extend the productivity cycle.

Vining peas

Of the two main commercial types of peas, those for fresh harvest, freezing or canning (known as vining peas) (Fig. 1.2) are produced in most of the world's temperate agricultural areas. The main constraints to production are weather, soil type and the availability of processing factories. The largest European producers of vining peas for freezing are the UK with 155,000 t and France with 190,000 t, Belgium with 69,000 t and Spain with 62,000 t produced annually. The USA is also a large producer with around 260,000 t per annum (Fig. 1.3).

Peas harvested fresh as immature pods and seeds are known as garden or fresh peas and are usually harvested by hand. An additional small quantity is grown for whole-pod consumption and the crop is known as mangetout, sugar

Fig. 1.2. Vining peas.

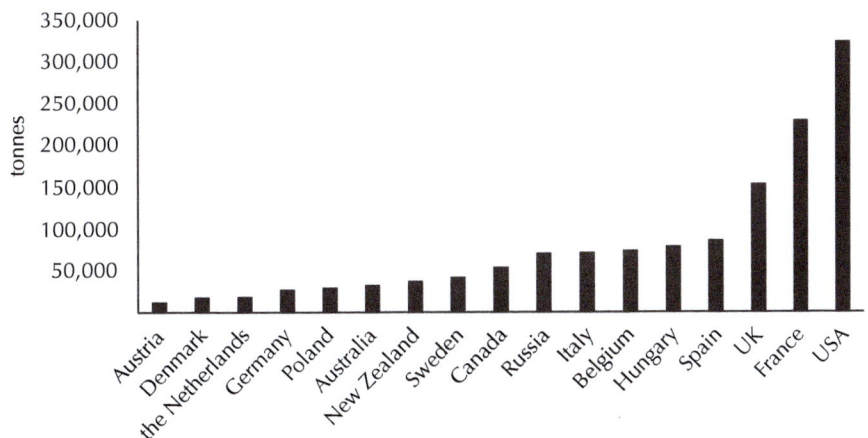

Fig. 1.3. Major producers of peas for processing (from FAOSTAT, 2013).

snaps or snow peas. However, the main production of peas as a 'fresh' vegetable is as vining peas, where the pods are mechanically removed from the stems, threshed green in the field and the peas are processed within a few hours of harvest as either frozen or canned peas.

Since the development of canning and freezing techniques, peas have been considered an important vegetable for use on their own or in mixed vegetable packs or 'ready meals'.

In the USA, peas were first brought over by European immigrants in the 16th century. From New England, settlers moving by wagon train brought the crop first to Wisconsin, where the canning industry would begin, and thence to the west, where they are now grown in Idaho, Washington and Oregon as both dried and green peas. The first steps to preserving fresh peas came in 1885 at the Paris Exhibition when a Madame Faure exhibited a hand-operated machine that shelled the peas from their green pods. The principles of this pea viner are still used today in self-propelled pea vining machines that operate in the field and are capable of harvesting many tonnes of peas per day.

The next advance was the development of preservation by canning; and the vegetable canning industry began in the 19th century.

Frozen peas began to be processed in the early 1920s primarily by fish freezers, hence the development of freezing factories in major fishing ports. The major growth in frozen peas was in the 1950s to 1970s with the increase in consumption linked to demands for convenience foods and the ownership of home freezers. The amount of peas grown for canning has declined since the early 1930s and today around 70% of vining peas are destined for quick freezing.

The quantity of peas frozen in recent years has begun to fall as consumption of vegetables has also dropped in developed countries. In the UK the tonnage of frozen peas has fallen from 155,000 t to around 140,000 t over the past 5 years and this trend also appears to be occurring in the USA and Europe.

Harvesting

Pea varieties continue to show improvements through careful breeding. Peas were at first pale in colour and susceptible to poor weather conditions and disease. Modern varieties are very tolerant to disease and are high yielding with generally improved flavour. The plant architecture has been changed in many varieties by reducing the plant height, reducing the vegetative growth by modifying the leaf shape or replacing leaves with tendrils and strengthened stems. A range of types are available to provide a 6–8-week harvest period allowing a continuity of fresh product to the factory for freezing or canning.

Harvesting is now mechanized on a large scale. By the 1950s farm- and factory-owned static viners were present on many farms. The use of these was based on peas being cut in the field, loaded on to a trailer and transported to the farm. The peas were then fed by hand on elevators to the static viners, where the peas were threshed, washed and chilled before transport to the factory. The next development was the introduction of trailed viners, which picked up a cut crop in the field and vined the peas *in situ*, unloading the vined peas into a trailer before transporting to the factory.

Since then, complete pea harvesters are the main form of harvesting with vined peas being delivered to the factory within 2 h of harvesting. Control of harvesting is now based on the accurate measurement of maturity. In the early 1950s, the Martin pea tenderometer was invented and even today derivatives of the original tenderometer provide the farmer and factory with a robust means of assessing the optimum time of harvesting.

The freezing process has also been modernized, with most of the processes and quality control being carried out automatically. Because of the necessity for a processing factory to operate at maximum efficiency over the 6–8-week harvest period, it is usual to have fixed tonnages of peas delivered over each 24 h period. In order to supply the required amounts, growers in the UK and in some other countries have formed pea cooperatives that may jointly own all the growing and harvesting equipment, organize the programme of seed sowing to allow continuity of supply and work together as a single group in harvesting and transporting to the factories. Currently in the UK there are around 12 pea cooperatives supplying 130,000 t of peas to eight freezing factories and one canning factory throughout the season.

Fresh peas

Production

Fresh peas are harvested as whole pods and are harvested by hand to minimize pod damage. The peas are stripped from the stems in the field and packed in boxes before being transported to the farm or pack house for packing.

Three types of fresh peas are marketed, including the standard garden pea or English pea varieties, where the pods are picked for hand shelling after sale, the sugar snap and the snow peas or mangetout types, which are picked when the seeds have developed but before the pod becomes stringy. Garden or English peas are usually grown on a large scale in fields with no additional support for the stems, in a similar way to vining peas, whereas the other two types may be grown in beds or on the flat. The variety chosen will dictate if they are grown on the flat or if support by some form of wire or trellis system is required. Plants grown on the flat are usually harvested at the same time, whereas crops grown on wire or trellis systems will be harvested over a period of time allowing several harvest dates per crop. Some crops are planted in the autumn and protected by a fleece covering, which is removed the following spring once the frost risk has passed. This system allows an early harvest. A large degree of hand labour is required for these crops but the final product can demand a premium price. Garden peas are more suited to cool temperate climates and can be grown on a range of medium to light soil types, but the whole-podded types are usually grown in areas where the temperature variation is not great and freedom from the risk of frosts or excessive rainfall is guaranteed. Pod quality is essential for achieving customer purchase and pods can be easily damaged by weather extremes. Edible podded varieties are generally more prone to root rot issues than garden peas.

Dry peas (combining peas)

Production

Dry-harvest peas are more commonly grown on a large scale commercially in Europe, particularly in France. Most of these are grown for the animal feed market. By comparison, production in the UK is smaller but the peas are grown for high-quality human consumption and export markets. The USA and Canada together are very large producers of dried peas, again mostly for animal feed though a significant amount is used as food ingredients (Fig. 1.4).

The largest producing country is Canada at 2.5 million tonnes; China and the Russian Federation each average around 1 million tonnes, whilst France, India and the USA each produce 0.5 million tonnes of dry-harvest peas annually (Fig. 1.5).

There are many uses for the dried pea seed, including animal feed and for human consumption. Peas have a high starch and protein content and therefore are suitable for a wide range of applications, whether whole, split or milled.

Because of their relatively short growing season, most spring-sown crops of combining peas have fully matured 5–6 months after sowing and the harvesting season usually coincides with dry weather conditions, allowing large-scale field harvesting to take place over a very short period. This is also essential because mature peas are susceptible to various losses, including by shelling out when the dehiscent pods become excessively dry. In large-scale operations, harvesting is mechanized and utilizes the same equipment as does

Fig. 1.4. Combining peas.

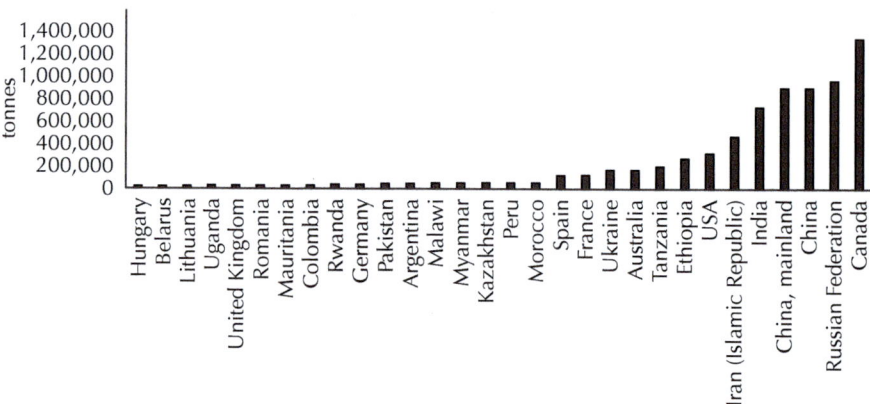

Fig. 1.5. Major producers of combining peas (from FAOSTAT, 2013).

cereal harvesting. Time of harvest is often just prior to that of winter wheat but before that of faba beans. As a significant proportion of the world crops are used for human consumption, there is a demand for good-quality seeds and even where peas are used for animal feed the quality of the produce may affect the purchase price. Peas can be affected by unfavourable weather conditions at harvest. Staining of the produce by saprophytic moulds can occur if the crop has been lodged during a period of wet weather and bleaching can occur caused by alternating rain showers and bright sunshine. The recent introduction of stiffer-stemmed varieties has improved the ability of peas to withstand some adverse conditions, but further improvement is desired.

Harvesting

Combining peas are normally direct harvested using a cereal combine when the crop has completely matured. In some situations, a chemical desiccant is used prior to harvest. In smaller-scale production, the crop can be cut and left in windrows to complete the drying process before being threshed. Harvesting is best carried out when the peas pass through the harvester without damage, when the moisture content of the peas has fallen to 25% or below. They may then be dried artificially. Damage in the form of seed splitting or seed coat cracking can occur at low moisture contents. Peas harvested at this slightly higher moisture content usually retain their colour even after drying. Peas for animal feed may be harvested at a lower moisture level, thereby reducing the need for artificial drying.

There are several adjustments to the combine harvester to handle peas, including reduction of the threshing drum speed and fitting appropriate screens to prevent peas being carried out with the straw.

For long-term storage, peas should be dried to 14% moisture content. Drying may be carried out artificially but the seeds are relatively large and extracting moisture from the centre can be a slow process compared with drying a cereal crop. They also have a low resistance to air flow and when drying in bulk there is very little sideways movement of air. Ideally, the temperature of blown air should be around 43°C or seed coat damage can occur. When the crop is intended for use as seed, germination may be adversely affected by high temperatures and too rapid drying.

Pea haulm is useful as an animal feed and it is often collected and baled immediately after harvest. The nutritional value of the straw is similar to that of barley straw but slightly higher in protein and lower in fibre.

VICIA FABA

Faba beans (field beans)

History

Vicia faba is known in Europe as field beans and is one of the oldest cultivated crops, thought to have originated in west or central Asia. *V. faba* is known to have been growing in the Near East and the Mediterranean area and grown as a food crop since 6000 BC. From the Near East the crop may have spread to Central Europe and Russia through Anatolia, the Danube valley and the Caucasus and from the Mediterranean coast to Egypt and the Arabian coast. It spread through Abyssinia and through Mesopotamia to India and China probably during the first millennium AD. It is essentially an Old World crop but was introduced to America in the 16th century and by the late 20th century it had reached Australia as a commercial crop. There are frequent references to beans by the ancient Egyptians, Greeks and Romans, with one of the oldest archaeological remains of seeds being found in Nazareth and dated between 6500 and 6000 BC. The first crop remains in the UK were identified during the excavation of an Iron Age site at Glastonbury.

Faba beans are now widely grown in Europe and in some parts of southeast Australia, the Middle East (including Egypt) and a small amount in Canada. There is a very large area of beans grown in China, mainly for domestic use. It is regarded as a temperate crop species, unsuited to the tropics other than at high altitudes: it was taken to the Andes by the Spaniards (Fig. 1.6).

Faba beans are still widely grown in the Mediterranean region, despite a proportion of the population having an allergy to the crop known as favism (Zinkham *et al.*, 1958). This condition is now known as an inherited deficiency among some people of the Mediterranean lacking the enzyme glucose-6-phosphate dehydrogenase in their red blood cells. Eating raw beans or breathing in pollen causes sudden destruction of red blood cells or acute haemolytic anaemia. There has been much conjecture on the association of favism and the death of Pythagoras in ancient Greece. Stories have been recounted where Pythagoras would not attempt to escape from his enemies by entering a field of beans, but the condition of favism is so rare it would seem unlikely that this is the reason for his refusal to go near to the crop (Simoons, 1996). Later in the

Fig. 1.6. Faba beans.

1st century AD the oldest surviving cookery book by the Roman, Apicius, gave nine recipes for beans with both mature (dried field) and immature (broad) beans being used (Edwards, 1985).

Notes on husbandry also survive from the classical period and Pliny recorded that 'when the bean is in flower, it requires water, but when it has passed blossom, it requires little' (Bostock, 1828). In the Middle Ages interest in beans was probably stimulated by reports of medical properties. In addition to their use in cures for 'old pains contusions and wounds of the sinews, the sciatica and gout,' Culpeper describes the use of beans for 'clearing the face of spots and wrinkles' (Sibley, 1802). Such comments may have helped to increase production. However, the use of field bean seed as a source of high protein and carbohydrate meant that they were used both for human and animal nutrition.

Beans were grown on a wide scale in northern Europe at the end of the 19th century and in 1873 in the UK alone about 224,000 ha was grown (primarily for horse feed), an area equivalent to that of wheat. One of the major reasons for the decline was increasing availability of cheaper protein from abroad. There was a slight revival during the First and Second World Wars because of importation difficulties. There was also an increase in livestock production and most of the beans were used as animal feed. With the development of the internal combustion engine and the introduction of machinery on to farms, the need for a home-grown supply of beans for farm horses declined; and from the end of the Second World War, imports of soya bean as a cheap protein feed for livestock increased dramatically. However, their value as profitable low-input soil-improving break crops in mainly cereal rotations that give produce that can be used as an alternative to imported soybean meal has meant that there is now more interest in the crop and so field beans have had a revival, particularly in northern Europe and Australasia.

Production

Internationally, the term 'faba bean' is commonly used to avoid confusion with other types of beans, some of which are also described as field or common beans. The latter are usually placed in the group of *Phaseolus* types, which will be described later. 'Faba bean' is therefore a useful generic term for both field beans and broad beans that botanically belong to the species *Vicia faba* L. (partim). 'Broad beans' is the term given to those varieties of *V. faba* that are harvested before the seeds have matured and dried and are used as a vegetable.

Although there are other members of the genus *Vicia* grown as crop plants, such as *Vicia sativa* (the common vetch), *V. faba* is genetically remote and no successful interspecific crosses have been made. This has limited the variability available for the breeding of *V. faba* and the development of varieties has been made even more difficult as a result of frequent cross-pollination between the different varieties, resulting in a wide range of genetic characteristics within a single population.

Faba beans are generally classed as winter (autumn-sown) varieties or spring (spring-sown) beans. Winter beans are generally more suited to soils that are cold and heavy and difficult to cultivate in early spring. There are more opportunities to plant winter beans after a dry summer into favourable soil conditions, thereby allowing them to become established as young plants before the onset of winter. Varieties are generally of high vigour to allow rapid germination and emergence in a wide range of conditions. The plants have ample opportunity to produce several production shoots, which help to maintain the plant height to a manageable level whilst compensating for a lower plant population.

Spring varieties of faba beans tend to have a wider diversity of varietal characteristics with some being relatively shorter, slightly earlier maturing and easier to harvest mechanically. This has allowed the cultivation of spring faba beans to develop in more northerly temperate areas, as well as widening the production opportunities in a number of European and Australasian countries. However, the performance of spring beans is adversely affected by water stress and drought, particularly during the flowering and reproductive stages of growth.

In agricultural terms the development of autumn-sown and spring-sown varieties of faba bean has led to major differences in varieties, methods of establishment, pest and disease profiles and date and ease of harvesting and so in many respects faba beans can be regarded as different crops. This increases the value of the species to farming systems in many temperate areas and allows consideration as to the type of crop that is most suitable for specific growing conditions.

Faba beans are one of the world's most important legume crop and world production exceeds 4 million tonnes, though production is still insufficient for human consumption in developing countries. Currently, the largest areas of faba bean production are in China, where around 1 million hectares are grown, the UK with around 100,000 ha and production of 400,000t per annum, followed by France with 80,000 ha, giving around 300,000t, whilst Australia produces around 150,000 ha and 300,000t (Fig. 1.7).

Within both types there is a range of varietal characteristics, including white- and coloured-flowering varieties. The white-flowered varieties are very low in tannin, which is perceived as an anti-nutritive factor in certain livestock diets. Coloured-flower types are easier to grow and the production of white-flowered beans is much more difficult to achieve because of their ability to cross-pollinate. Because of their versatility in an arable rotation, beans are popular in that they provide a useful non-cereal break crop, can reduce the pressure on seasonal workload and have a ready market both for human consumption and for livestock, including pork, beef, dairy and aquaculture production.

Autumn-planted beans (winter beans) can be drilled either in non-inversion tillage or after ploughing. Seed is planted at lower density than those

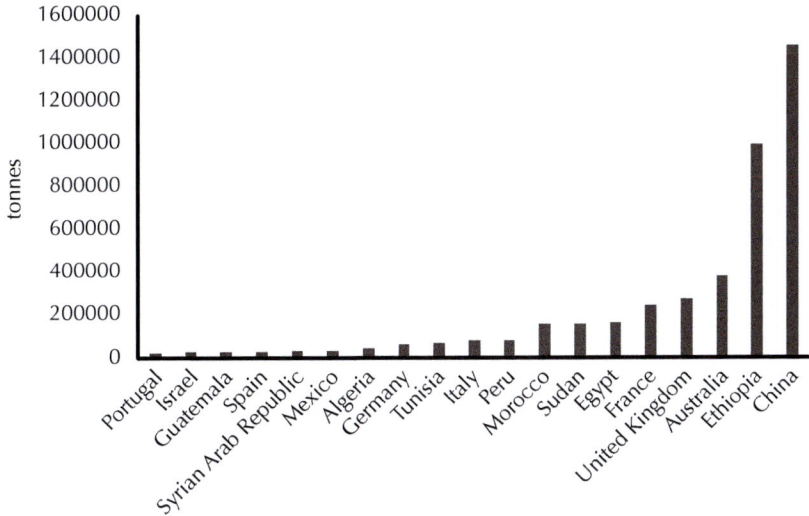

Fig. 1.7. Major producers of faba beans (from FAOSTAT, 2013).

grown in the spring, as the cool temperatures of winter encourage the plant to produce side shoots that later develop into fertile stems, whereas spring-planted beans tend to produce only one or two side shoots.

Autumn-planted beans are usually grown on heavier wetter land where it is not possible to produce a reasonable seedbed for spring crops. Winter beans may also suffer from bird or frost damage, though they can withstand temperatures down to −12°C and lower if there is snow cover. Current varieties tend to have a large seed and therefore conventional seed drills for cereals are not always suitable; however, modified direct seeders and deep-tined drills are used successfully in the UK and France where winter beans are mainly grown. Winter beans may also become infected with leaf spots such as chocolate spot (*Botrytis fabae*) and leaf and pod spot (*Ascochyta fabae*).

Some taller varieties may lodge in severe weather, but modern varieties are less indeterminate and stem strength is improved. Spring-sown beans (spring beans) are less likely to suffer from chocolate spot but they can be more susceptible to downy mildew (*Peronospora viciae*) and to aphids and aphid-transmitted viruses.

At whatever time of the year beans are planted, both winter and spring varieties tend to mature at the same time and therefore only in extreme conditions will the harvest be delayed beyond the autumn period, thereby allowing the completion of the cereal harvest before the beans are ready.

Harvesting

Harvesting is by direct combining with or without the need for chemical desiccation. Faba bean pods blacken and seeds become dry and hard before the stems of most of the plants. The pods are easily threshed when the moisture content is 16–20% but it is usual to wait for the majority of the stems in the crop to become brown and dry, to allow an easier combining operation.

In very dry seasons, when beans are harvested at very low moisture contents, seed coat cracking or seed splitting can occur. Whilst this is not too serious where the beans are intended for milling as animal feed, beans for the high-quality export market for human consumption should be handled much more carefully and harvested at a higher moisture content to avoid damage. The danger of damage to very dry crops is also important for seed production, as the germination capacity of damaged seed is reduced. Because of the large seed size, the drum speed of the harvester must be at its lowest setting and the concave clearance must be at the maximum. Once harvested, the beans can be dried using ambient or slightly warmed air until the optimum storage condition of 85% dry matter of the beans has been achieved. Dry bean haulm can be used as a feed for livestock and can be made into bales for storage after the crop has been harvested.

Broad beans

Broad bean crops tend to be mostly spring-sown varieties grown for harvesting as a vegetable, shelled and either frozen or canned, or harvested as whole pods for the fresh market. Most broad beans are therefore grown on a large scale and harvested mechanically, ensuring supply to the processing factory, or they may be grown on a smaller scale for hand harvesting of the whole pods (Fig. 1.8).

Production

As with vining peas, broad beans are either grown for the canning or freezing market or picked as fresh pods and sold as whole podded beans for the fresh market. There are many similarities with vining peas when broad beans are grown for processing. Crops are drilled in the spring in northern Europe, but in the warmer areas crops are established in the autumn in order for them to establish an extensive root system to withstand summer drought. Planting dates are regulated by the projected harvest date. In the UK and Europe, broad bean harvesting is planned to follow the pea harvest although that is not always the optimum timing to achieve maximum yield, as yield tends to decrease with a later planting date. A range of varieties can be used to complete a harvesting programme, though the number of varieties suitable for mechanical harvesting and to provide a green product after processing is limited: only white-flowered varieties are used, because the anthocyanins produced in the seed of coloured-flowered varieties discolour the beans during the heating and cooking process.

Broad beans can be grown successfully in many temperate areas, including Europe, the USA and Australasia. In general, broad beans are grown in

Fig. 1.8. Broad beans.

conjunction with vining peas as the same harvesting equipment is used. Similarly the processing factories produce both final products.

Beans require adequate moisture to ensure good establishment and they are more susceptible to drought stress than peas and so are better suited to the more moisture-retentive soil types. Beans can also suffer from excessive moisture and this can lead to tall crops with dense canopies that result in poor pod and seed set. The number of varieties is limited, because there are very few commercial breeders of broad beans, but there are differences in crop height, seed size and colour as well as some early-maturing varieties being available.

PHASEOLUS BEANS

It has been estimated that around 55 species of *Phaseolus* exist today (Debouk, 1991), all of which are American in origin. Within these species are the common bean (*Phaseolus vulgaris*) and the vegetable climbing bean (*Phaseolus coccineus*). The most commonly produced bean is that of the *P. vulgaris* family and there are many types of bean and many varieties with a wide range of botanical characteristics.

History
It had been believed that *P. vulgaris* originated from Asia but it has since been shown that beans were first domesticated in the New World, with

archaeological remains found first in Peru and later in south-west USA. Since then, several additional remains have been uncovered in the Andes, Middle America and North America. The beans evolved from a wild-growing vine into a widely grown agricultural crop. A detailed history of the origins and dispersal of *Phaseolus* was described by Gepts and Debouk (1991).

Production

A large proportion of the crop is now grown worldwide as a dry harvested bean that is commonly known as a dried bean, but a smaller proportion is also grown as a vegetable where the pods are harvested and consumed in the fresh state or preserved as a frozen product.

With these different forms of final produce, there is a huge variation in seed and pod characteristics in shape, size and colour and this has led to a wide range of varieties being developed and grown commercially. A number of varieties still retain their climbing characteristics but many have been developed as short upright bushy plants (Fig. 1.9).

Dried beans are now grown in many subtropical and more temperate regions but the most important areas are South and Central America and Africa, West Asia, Europe and the USA. FAO statistics (Fig. 1.10) indicate that Brazil and Argentina, Mexico, India, China and Myanmar (Burma), the USA and Canada produce the largest amounts of beans and the annual world total production is over 23 million tonnes.

Fig. 1.9.   Dried beans.

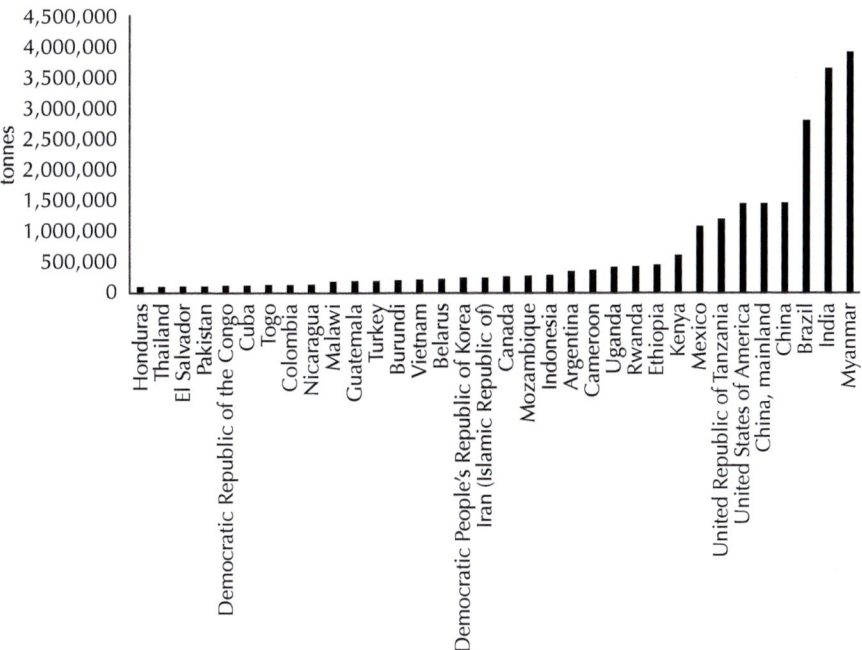

Fig. 1.10. Major producers of dried beans (from FAOSTAT, 2012).

Most bean varieties have been chosen or developed for speciality markets and consumers tend to prefer specific types. Because of this there is a very wide range of characteristics, particularly seed size and colour, available as varieties for these specific requirements. Most dried beans are rehydrated and cooked either whole or as a bean flour or porridge. Dried beans can be stored successfully and are therefore a valuable source of protein in the human diet in developing countries. They are easy to transport and process industrially. A large quantity of white-seeded dried beans are canned in tomato sauce and sold as baked beans in the USA, Canada and the UK, whilst colour-seeded beans are canned or processed for bean salads or ready meals.

Green beans (dwarf, French or snap beans)

Beans may be described in groups depending on their maturity when they are harvested. Green, French or snap beans are grown as horticultural crops and harvested fresh; they are consumed as pods, either fresh or processed (Fig. 1.11).

Green beans are an important crop in developed countries and are grown on a large commercial scale in Europe and the USA. The range of varieties available is extensive, often characterized by pod length, width and colour. Although most of the commercially grown varieties are green podded,

Fig. 1.11. Green beans.

variants of colour from dark green to yellow are available; yellow-podded varieties are known as wax beans. Additionally the pod shape can vary from round cross-section (Blue Lake types) to flat cross-section (Romano types).

Production

The crop is harvested at the whole-pod stage for freezing or canning. Both canners and freezers present beans in whole, cut or sliced forms. In each case the processed product should be of uniform colour, free from unpleasant odour and of reasonably firm and fibreless texture. Beans intended for whole pods are usually of small diameter and a length of about 10–12 cm, whereas beans for slicing or cutting tend to be wider in diameter. The broader flatter pods (Romano types) are not generally used for processing.

Harvesting

For the fresh market, beans are still mainly hand picked. Several attempts at using a mobile bean-harvesting machine have been tried but the damage caused to the pods is considered as unacceptably high for the fresh market pack. Beans are grown in a similar way to fresh-market peas but with a greater row width to allow access for the picking team. Beans are packed in boxes and transported to the pack house for sorting, repacking and cooling. Varieties with very thin pods are known as fine beans and imports of these into Europe are usually from African countries, particularly Kenya. Fine beans are picked very carefully to minimize handling damage to the pods. Blemishes to the pod commonly occur during the picking and packing operation; and once the boxes have been covered with plastic film, fungal deterioration can be rapid at high relative humidity.

SUMMARY

Peas and beans have been domesticated and produced on a large scale in many parts of the world over a long period of time. There have been significant advances in production and harvesting techniques which have enabled large-scale production both in area and quantity. In general, demand is increasing and peas and beans are being traded by major production countries worldwide.

REFERENCES

Bostock, J. (1828) *The First and Thirty-third Book of Pliny's Natural History*. Translated by J. Bostock. Baldwin and Cradock, London.

Debouk, D. (1991) Systematics and morphology. In: van Schoonhoven, A. and Voysest, O. (eds) *Common Beans – Research for Crop Improvement*. CAB International, Wallingford, UK, pp. 55–118.

Edwards, J. (1985) *The Roman Cooking of Apicius*. Rider and Co., London.

Gepts, P. and Debouk, D. (1991) Origin, domestication and evolution of the common bean (*Phaseolus vulgaris*) In: van Schoonhoven, A. and Voysest, O. (eds) *Common Beans – Research for Crop Improvement*. CAB International, Wallingford, UK, pp. 7–54.

Group, L.P.W. (2013) Legume phylogeny and classification in the 21st century: progress, prospects and lessons for other species-rich clades. *Taxonomy* 62, 217–248.

Lewis, G.P., Schrire, B., Mackinder, B. and Lock, M. (2005) *Legumes of the World*. Royal Botanic Gardens, Kew, Richmond, UK.

Sibley, E. (1802) *Culpeper's English Physician and Complete Herbal*. Oxford University, Oxford, UK.

Simoons, F.J. (1996) *Plants of Life, Plants of Death*. University of Wisconsin Press, Madison, Wisconsin, p. 216.

Zinkham, W., Lenhard Jr, R.E. and Childs, B. (1958) A deficiency of glucose-6-phosphate dehydrogenase activity in erythrocytes from patients with favism. *Bulletin of the Johns Hopkins Hospital* 102, 169.

BOTANY AND PHYSIOLOGY

LEGUMES AND NITROGEN

Nitrogen is a major element in crop nutrition and in the absence of nitrogen, crop growth is severely affected as a result of a general chlorosis and reduction in photosynthetic ability. In most agricultural systems, nitrogen (N) is applied in a readily available form, either as nitrate in a chemical fertilizer or as manure or compost, where microbial breakdown can release soluble forms of nitrate that are then taken up by the growing crop. In large-scale commercial agriculture, most nitrogen is applied as a fertilizer produced by a chemical process (the Haber–Bosch process) that converts nitrogen from the atmosphere into ammonia using high quantities of energy, or from non-renewable mined sources of minerals. In whichever production method is used, there is a high economic cost involved. An additional problem occurs with the use of applied fertilizers when excess chemical is leached out of the soil by rainfall or irrigation and is then able to enter water courses and catchments.

In natural situations where there are low levels of mineral nitrogen in the soil, legumes are particularly successful due to their ability to fix atmospheric nitrogen into a form that they can utilize for normal growth. A number of other plants that have this characteristic, such as species of *Gunnera*, *Alnus*, *Casuarina* and *Mirica* have a symbiotic relationship with soil-borne organisms that colonize the roots. The association of plants and beneficial organisms has been shown by fossil records to have been a major factor in the success of plant colonization. Although the genes required for signalling the onset of root colonization have been characterized, the origins of these genes remain elusive (Delaux *et al.*, 2015).

Legumes have a specific relationship with soil bacteria of the *Rhizobium* genus where bacteria invade roots, leading to a process whereby the conversion of atmospheric to soluble nitrogen occurs. The trait of symbiosis with *Rhizobium* is one of the main reasons why legumes are characterized as 'pioneer plants', as they are commonly found to colonize soils of low N content, and often in bare and exposed environments (Cloutier *et al.*, 1996). Crops that have this ability to form symbiotic relationships are one of the main sources of nitrogen in agricultural systems that have limited inputs. Legumes are therefore very economically important crops.

Biological nitrogen fixation

Although there are other means whereby atmospheric nitrogen is converted to nitrate, e.g. during electrical storms, the process whereby plants are able to fix their own nitrogen is referred to as biological nitrogen fixation (BNF). The details of BNF have been described by Howard and Rees (1996) and many others. Rhizobia, which are present in many soils of the world, are stimulated to infect via the root hairs or through cracks in the epidermis of host plants by root exudates containing flavonoids that are released into the soil during root growth (Sprent and James, 2007). The genetic mechanisms involved in initiating this infection are currently being examined. The presence of calcium and its associated signalling within the root hairs have been shown to be an important factor in the whole process. Nitrogen-fixing rhizobial bacteria that associate with legumes signal the start of symbiosis to their hosts through the release of diffusible compounds including lipo-chito-oligosaccharides which are also known as Nod-factors. These cause oscillations of calcium ions in the root epidermal cells (Sun *et al.*, 2015). The basic elements of this signalling being elucidated by current research will become a cornerstone for exploring the possibility of engineering similar symbiotic associations with non-legume crops (Oldroyd and Dixon, 2014).

Once established in the roots, the bacteria multiply and move to the root cortex via an infection thread, which is a tube made from the walls of the cells. The roots react to this infection by producing root nodules, which occur as a result of the enlargement of the cells of the root tissue (Fig. 2.1).

The movement of the bacteria within the roots results in the production of new nodules and as the nodules age they senesce and degrade, releasing rhizobia into the soil to invade new roots. Nodules provide a specialized niche to allow the rhizobia to fix nitrogen in conditions of minimal air, which protects the enzyme nitrogenase but allows enough oxygen to be present to enable aerobic respiration associated with nitrogen fixation (Gallon, 1992; Minchin *et al.*, 2008). The nodule also allows the symbiotic bacteria to efficiently exchange their fixed nitrogen in return for carbohydrates from the host plant. Nitrogen is converted first to ammonia, using energy supplied by the metabolism of carbohydrates (Emerich and Burris, 1978). The ammonia is then converted into amino acids that are utilized by the bacteria to produce proteins and peptones for their growth. The fixed nitrogen incorporated into the rhizobia is then released as amino acids, which are taken up by the host plant (White *et al.*, 2007; Anderson *et al.*, 2013).

Fig. 2.1. Nodules on faba bean roots.

Nitrogenase is very sensitive to oxygen but rhizobia require oxygen for respiration. Nodules contain high concentrations of leghaemoglobin, which carries oxygen and surrounds the bacteria in the infected nodules. This leghaemoglobin is able to transfer a low concentration of oxygen to the bacteria and therefore its presence in nodules is essential. In anaerobic conditions, or in waterlogged soils, the rate of nitrogen fixation is severely reduced and a well aerated soil type is necessary for efficient biological nitrogen fixation to occur.

Not all legume species have a relationship with the same strain or species of soil-borne *Rhizobium*. Nodulation will not occur if the associated species is not present in the soil. Within each species of *Rhizobium*, there are a number of host-specific strains. Typically, *Vicia* and *Pisum* spp. are associated with *Rhizobium leguminosarum* but it is thought that, within that species, different strains of bacteria are favoured by each crop type. Recent studies have also indicated that there are two distinct genetic classes of *Rhizobium* that nodulate *Vicia faba* (Del Egido, 2014). *Phaseolus vulgaris* is associated with *Rhizobium phaseoli*; and lupin species, as another example, have a relationship with *Bradyrhizobium lupini*.

In many cases, the *Rhizobium* species/subspecies that infects particular host plants is present in soils but in some situations, especially where non-native crops are grown, seeds can be inoculated with commercial preparations of the specific strains or species of *Rhizobium* for that crop (Table 2.1).

The amount of nitrogen fixed depends on many factors, such as the effectiveness of the *Rhizobium* strain, the condition of the host plant, physical factors of the environment and the agronomic practices. The amount fixed varies considerably; it has been variously reported as between 45 kg and 550 kg N/ha for *V. faba* as an example (Nutman, 1976) and has been compared for a range of legume species by Salunkhe and Deshpande (1991) (Table 2.2).

Efficiency is also affected by the nitrogen status of the soil. This is especially obvious where soil levels of inorganic nitrogen are high and this suppresses natural nodulation and nitrogen fixation (Slattery *et al.*, 2004). Temperature also affects the rate of nitrogen fixation and extremes of high or low temperatures reduce efficiency. Nitrogen fixation in legumes is sensitive to excess water (James and Crawford, 1998), drought and the salinity level of the soil coupled with high pH. A shortage of key minerals may also limit BNF, particularly phosphorus (P). In *Medicago* species, nodule size and weight were reduced where P was deficient (Schultze and Drevon, 2005).

Table 2.1. *Rhizobium* species and their host range (compiled from Smartt, 1976).

Microsymbiont	Host affinity	Crop
Rhizobium japonicum	Glycine	Soybean
Rhizobium leguminosarum	*Pisum, Vicia, Lens, Lathyrus*	Pea, faba bean, lentil, sweet pea, grasspea
Bradyrhizobium lupini	*Lupinus*	Lupin
Rhizobium meliloti	*Melilota, Medicago, Trigonella*	Lucerne, sweet clover, alfalfa, fenugreek
Rhizobium phaseoli	*Phaseolus*	Common bean
Rhizobium trifolii	*Trifolium*	Clover

Table 2.2. Relative nitrogen fixation of some legume crops (Salunkhe and Deshpande, 1991).

Legume crop	Nitrogen fixed kg/ha/year
Alfalfa (lucerne)	114–223
Clover	21–36
Chickpea	24–84
Phaseolus	178–251
Vicia faba	174–196
Pea	87–222
Lentil	167–189
Lupin	121–157
Soybean	22–310

Nitrogen fixation is greatest during the periods of flowering and pod filling, when the light intensity is greatest during the spring and summer months and, in most healthy crops of peas and beans, the amount of nitrogen fixed is adequate for the requirements of the crop throughout its life. In the initial growth stages, nitrogen is used in stem and leaf development; and later, during seed development, much is stored in the maturing seed as protein.

After harvest, when much of the fixed nitrogen as protein is removed from the field, there is a residual level of nitrogen in the haulm and in the root system and nodules. Where roots are left to degrade in the soil, nitrate is released slowly and is available for uptake by the following crop. These levels are likely to vary considerably but are recognized as being around 20–50 kg N/ha (Sylvester-Bradley and Cross, 1991), as confirmed by Ward and Palmer (2013). For faba beans, estimates in the UK suggest fixation of 60–100 kg N/ha within the stem and root remains after harvest (Iannetta et al., 2015). Although there is little information on the amount of nitrogen produced by legume crops, there have been estimates that globally 50–140 million tonnes of nitrogen is fixed each year by BNF (Unkovich et al., 2008). More recently, Baddeley et al. (2013) calculated that around 0.8 million tonnes of fixed nitrogen occurred within the EU in 2009.

This source of nitrogen is economically valuable in the cropping system, as it provides a source of nitrogen immediately for the following crop as this becomes established. Since the fixed nitrogen is released relatively slowly in the soil, there is less of a risk of leaching during wet periods than with applied fertilizers but, equally importantly, the residual nitrogen available for the following crop reduces the requirement of the latter by around 50 kg/ha. In addition, the inclusion of a legume pulse crop in an arable rotation provides a 'break' in a predominantly cereal-based system, reducing the build-up of cereal pathogens such as take-all (Gaeumannomyces graminis) and providing an opportunity to control perennial weeds and grassweeds, such as black-grass (Alopecurus myosuroides), which has developed resistance to commonly used graminicides. In the EU, it has been calculated that the yield of winter wheat immediately following a pulse crop is increased by 0.6 and 0.9 t/ha (Von Richtofen, 2006) and similar responses have been noted in Canada (Wright, 1990). In some studies with V. faba, its inclusion in the crop rotation enhances the diversity of wild flora and fauna as well as soil microflora (Kopke and Nemecek, 2010). There is, therefore, compelling evidence for an arable rotation that is supported by the inclusion of a legume crop (Iannetta et al., 2013).

PEA PHYSIOLOGY

Germination

Seed germination begins soon after the onset of imbibition. The seed cotyledons comprise embryonic respiring tissue, as there is no separate nutrient storage organ in the seed. Once imbibition is complete, the starch present in the cells of the cotyledons begin to be converted to sucrose, which then becomes available as an energy source to the embryo axis. The embryo axis then elongates, ruptures the testa and becomes visible as the developing radicle or root. Later the apical part of the embryo develops as the plumule and extends upwards and emerges above the soil surface. The cotyledons remain below ground (hypogeal germination) where they continue to supply nutrients to the developing seedling until the plumule expands to form the primary set of leaves.

Vegetative growth

The pea plant is not particularly effective in supporting itself, as the stems are inherently weak. The ancient parents of modern pea varieties were largely prostrate, with an extensive stem growth. However, the ability of the pea plant to stand more erect is a major objective to breeders of modern varieties. There is usually one main stem from each plant but the number of pods per node and per plant depends on variety. In some cases, there may be one or more axillary stems, but the leaves and flowers have the same structural formation. As the stem develops, the leaf is produced from an axillary meristem known as a node. Leaves develop at each node from about the sixth node upwards. More leaves develop on the stem as it elongates and the space between the nodes is known as the internode. Large stipules are borne at each node. Each section of the stem consisting of the internode, leaf and meristem is known as a phytomer (Nougarède and Rondet, 1973). Each compound leaf consists of a petiole with four to six pairs of pinnate leaflets and terminating in three tendrils. The pea leaf has a cuticle of wax on the upper surface and leaf colour can range from yellow-green to deep blue-green, according to variety. Recently, the general morphology of the plant has been modified to improve agricultural adaptability and many modern varieties derived from selected mutants have different stipule, leaf and tendril characteristics. Types have been developed where the stipule is much reduced and the leaflets have all been converted to tendrils. These '*afila*' types have been referred to as leafless peas, but the productivity of these types is generally inferior to the conventionally leaved peas. Further selection work has developed varieties with a normal-sized stipule whilst retaining the *afila* characteristics (Figs 2.2 and 2.3). These 'semi-leafless' *afila* types are now represented by varieties of both dry harvesting peas and vining types. A number of varieties used commercially have normal or more pinnate leaflets.

Root development

Peas have a fine taproot that can penetrate to a depth of 80 cm, and less well developed lateral roots. Root growth follows soil fissures and has a poor ability

Fig. 2.2. Normal leaflets on a pea plant.

Fig. 2.3. *Afila* leaflets.

to penetrate consolidated conditions. As described above, as a legume the roots of a pea plant develop nodules that contain nitrogen-fixing bacteria (*Rhizobium leguminosarum*), providing the plant with sufficient nitrogen throughout its life. Nodules vary according to age and growth conditions. They are irregular in shape, measure approximately 2.0–5.0 mm in diameter and tend to be densely packed along the tap root. Whilst they are initially white, they discolour and senesce as the plant becomes mature. When young, healthy and active they are usually bright pink and, if cut open, a red coloration caused by the pigment in leghaemoglobin is visible. The bacteria benefit by having a stable environment and a supply of carbohydrates provided by the plant.

Peas are thus able to utilize enough nitrogen for their entire growth and also to provide a residual level of nitrogen in the soil after the crop has been removed, which is then available to a following agricultural crop. In temperate climates this is likely to be a cereal and the residual nitrogen substantially reduces the requirement for the application of a full rate of fertilizer nitrogen. This benefits an agricultural rotation that includes a legume, thereby reducing the overall nitrogen input into the system.

Flowering

Flower initiation is thought to be triggered by photoperiod and temperature but the number of nodes and the intrinsic earliness of the plant also determine at which node the first flowers are produced. In modern pea varieties, flowering begins on a pre-determined node, thereby providing a known time of maturity to aid and plan the harvesting operation. The number of flowering or reproductive nodes varies between genotype but in most spring-sown peas there are between six and eight flowering nodes that produce pods. In more indeterminate genotypes, the effects of environmental conditions can be more pronounced in that the stem continues to extend, producing more reproductive nodes. Plant density also influences stem length, as stem length and reproductive nodes can double where the plant density is very low.

There is a range of genotypes that exhibit different flower colours. Flowers, which are self-pollinated, may be white, pink, purple or bi-coloured, depending on variety. White-flowered peas usually produce green or yellow seeds, whereas coloured-flowered peas tend to produce brown or speckled seeds. Usually only white-flowered peas are grown for freezing or canning, though some markets exist for coloured-flowered varieties both for human consumption and for specialist pet foods.

Flower development progresses linearly along the stem as day temperatures accumulate and this is constant over a wide range of environmental conditions; however, during slow periods of growth due to low temperature, for example, the rate of progression may slow down. Onset of flowering is also affected by photoperiod in that the longer the day length experienced by the plants after emergence, the earlier the floral initiation will begin (Berry and Aitken, 1979) and this can also vary with the variety (Lejeune-Henault *et al.*, 1999).

Pea flowers contain both anthers and stigma and the pollen is mature before the flower opens. The flower consists of five petals, having one standard, two wings and two keels that are fused except at their base. Within the flower is the stamen, which is made up of the filament and the anthers, and the pistil, which contains the style, stigma and ovary. As the flower develops, the anthers are exposed to the stigma and pollen is transferred within the flower and self-pollination occurs before the petals are fully open. As the fertilized ovary expands, the walls elongate and the pod becomes exposed. The remains of the flower petals are carried on the tip of the pod until they desiccate and fall off.

Pod development

Modern varieties tend to have at least two pods per node, whilst some have multiple pods on each fertile node. Agronomic practices and environment will affect the pod setting and pod number characteristic, as high plant populations decrease the pod numbers over a given area. In general the pea pod is a dehiscent fruit, the sides being two carpels. When mature and dry, the pods have a natural tendency to split open and eject the seeds. Modern breeding attempts to limit this characteristic to reduce seed loss before harvest (Fig 2.4).

Fig. 2.4.   Dehiscent pea pod.

There is wide variation in the number of seeds per pod between varieties, though the number is generally five or six. Varieties grown for freezing or canning are sweet, having compound starch grains, and in the dry state the seeds are deeply wrinkled. Varieties grown for dry harvest are less sweet and have a higher starch content and the starch grains are simple. The cotyledons of the dry seed may be a blue-green colour, pale green, yellow or creamy white, olive, brown or reddish brown and may have speckled or marbled markings. Seeds also differ in size or shape and may be wrinkled, round or dimpled with a thousand-seed weight that ranges from 90 g to 400 g.

FABA BEAN

Germination

The seed germination process of faba beans is very similar to that of peas. With hypogeal germination, the embryo axis elongates and then ruptures the testa and the apical part of the embryo develops as the plumule emerging above the soil surface, with the seed remaining underground.

Vegetative development

Faba beans have a thick, square and slightly winged stem. This feature makes the stem stronger and less susceptible to lodging. Stem length varies greatly between cultivars but can be correlated with the internode distance, in that very short-stemmed cultivars have short internodes. The pattern of stem growth depends on cultivar and cultivation techniques but the branching habit of the plant becomes greater the earlier the sowing date. Thus in winter beans, which are sown in the autumn, the number of stem branches is higher than in spring-sown varieties. As plant density increases, the number of stem branches per plant decreases. The branching habit of faba beans is similar to that of most legumes, in possessing a centrifugally arranged series of buds initiated on each leaf axil (Dormer, 1954). Basal branching arises from these buds at the lower part of the stem (Bond and Poulsen, 1983).

Depending on variety and growing conditions, stem height varies from 120 cm to 200 cm but environmental conditions such as drought or excessive moisture can have a dramatic effect on crop height. In general, stem strength is good and allows the plant to remain erect until mature.

The first two leaves at the base of the stem remain vestigial, with only small stipules being visible. The first foliage leaf usually arises at the third node and the first few leaves have only two leaflets. Leaves are generally of two pairs of ovate leaflets, though the number of leaflet pairs increases once flower buds are initiated. There are toothed stipules but no tendrils.

Stem growth is affected by temperature as well as soil type, moisture avail-ability and incident light. During periods of low light intensity or in instances of high plant density, stem length increases as the internodes lengthen. Leaf growth is also affected by environmental conditions, including nutrient avail-ability and water stress.

Root development

Root growth of *V. faba* begins immediately after the seed starts its germination process. Within a week of germination a well established root cap develops; and after 2–3 days the primary root emerges from the seed and lateral root initiation begins. The primary root continues to grow downwards and lateral roots spread from the main tap root. Nodulation occurs as the root hairs are invaded by soil-borne *Rhizobium* bacteria and nodules develop on the upper part of the tap root, where they become abundant. Nodules tend to be smaller on the lateral roots but these smaller nodules take over the nitrogen fixation from the older larger ones on the tap root as the plant becomes older.

Flower development

Flowers are borne on short axillary racemes, with two to six flowers per raceme. Depending on variety, flowers may be completely white but both sepals and pet-als may have anthocyanin pigment ranging in intensity from purple to pink, and in most cases with black or purple melanin markings on the wing petals, but there are at least two genes that confer the absence of a wing spot and are pleitropically associated with the absence of colour on the standard petal, stem and stipule and absence of tannin in the seed coat (Fig. 2.5).

The structure of the flowers has become adapted to the insect pollinators, though some self-pollination occurs. They are characteristically papillionate, a zygomorphic form, and have longer corolla tubes than do most other legumes. The sepals are combined into a single five-toothed calyx. The irregular corolla is made up of five petals, the standard two wings and two lower petals that are united along their outer edge to form a keel. The flower has ten stamens, the upper one being physically free. The filaments of the other nine are united in a sheath which encloses the ovary. The single ovary has from two to nine ovules (occasionally more) arranged along the inner, upper suture. The style is almost at right angles to the ovary.

The petals are joined and hinged and tripping occurs when an insect depresses the wing and keel petals, releasing the stigma and pollen. At the same time, the pollen is transferred to the hairs of the visiting insect which is then able to cross-pollinate other flowers on the same or neighbouring plants. The flowers produce a strong scent, which is attractive to pollinators and other

Fig. 2.5. Coloured faba bean flower.

insects (Somerville, 2002). Bees play a decisive role in the pollination of beans. The importance of bees in cross-pollination and the improvement of its production is well known (Poulsen, 1975; Svendsen and Brødsgaard, 1997). The most frequent pollinators of *V. faba* flowers in temperate regions are bumblebees (*Bombus* spp.). In most countries, honey bees (*Apis mellifera*) collect pollen from *V. faba* and in some regions hives are often taken to crops at flowering time (Bond *et al.*, 1985).

The positions of the first flowers and first pods do not necessarily coincide, as flower abortion can occur immediately after the petal collapses and the number of flowers produced on a raceme and number of racemes per plant almost always exceeds the number of pods that develop to maturity. It is thought that around 10% of flowers survive to form mature pods. Environmental conditions have a large influence on this characteristic of faba beans, especially in conditions of low light intensity or shading in dense plant populations as the stem lengthens and produces more flowers on the upper parts of the stem. This may result in the lower flowers (which may even have been fertilized) falling off without setting pods.

Pod development

As the fertilized ovary develops as a pod, the remains of the petals senesce and fall off, leaving the pod exposed. The number of reproductive nodes with pods varies between varieties but more especially is influenced by environmental conditions. Pods are usually borne in groups of two or three at each node, beginning on the fourth or fifth leaf node and continuing up to the tenth or even higher nodes. Plant density and the number of branches influence the number of pod-bearing nodes in that plants of low population density produce a higher number than those growing at high density, provided that growth is not restricted by water availability, nutrient deficit or biotic stresses.

Pods are straight and fleshy and the internal surface of the pod is downy. The average number of seeds per pod ranges from two or fewer up to eight and there may be more than this in some varieties. It is usually a stable character, though there may be fewer seeds at upper nodes and those pods that are produced at the higher end of the stem can contain shrivelled ovules. The number of seeds may be higher in the larger-seeded varieties. Colour of the immature testa is usually cream or blue-grey to green, though the testa of mature seed is usually cream, tan, brown, purplish brown or black, but in most cases the colour darkens with age, particularly after harvest. Seed from faba bean with flowers that are not completely white contain anthocyanins that darken with age. White-flowered beans tend to produce a more grey-coloured mature seed colour. Hilum colour may be black or white and varies with variety. The preferred varieties of faba beans grown for human consumption of the dried produce usually produce a white hilum. Typical thousand-seed weights are 280–530 g but some broad bean varieties are significantly larger. Again, environmental conditions can affect seed weight and the most important factor is moisture availability. In dryer growing conditions, seed weight can be reduced by as much as 25%.

The majority of faba beans have dehiscent pods but are less likely to shatter before harvest than peas. Leaf senescence occurs during seed growth with declining root activity and increases with temperature and water shortage.

PHASEOLUS BEANS

The plant type has great variation in habit, response to photoperiod and in leaf, pod and seed characteristics, which vary considerably and so a simplified description of the morphology of *Phaseolus* was produced by Debouck (1991) to explain the main components of the plant.

Germination

Seed germination of *Phaseolus vulgaris*, and many other species, is epigeal in that the seed cotyledons emerge from the soil with the shoot. The first cotyledon leaves develop from the shoot as a single pair of opposite and (generally) ovate leaves. The remains of the cotyledon then shrivel and fall off the stem, leaving two small scars. The shoot continues to develop, producing one or more sets of pinnate and mostly trifoliate leaves. These are also known as true leaves.

Vegetative development

The stem continues to extend with more sets of trifoliate leaves on lateral axes which can vary in length and some branching can occur. The stems are round in cross-section and the lower part of the stem tends to be erect without torsion, whereas the upper part of the stem may begin to twist as in some climbing types around a support. However, the more determinate types of bean tend to remain upright and develop as bushy forms.

The leaves develop opposite and alternately up the stem but in the leaf axils buds develop which in turn produce secondary leaf branches. Leaves are trifoliate, with the two lateral leaflets being asymmetrical while the terminal leaflet is symmetrical. Leaves vary in colour from yellow-green to blue-green and may be slightly pubescent on the upper and lower surface.

Two distinctly different growth habits can be found among varieties of *Phaseolus* and these can be classified into two separate types: determinate and indeterminate. The difference is that in the former, main stem and lateral branches terminate in inflorescences, while in the latter, main stem and lateral branches are topped by a vegetative meristem capable of continuing to grow and develop more leaves and flowers. In these indeterminate varieties, flowers are located laterally, directly inserted at the nodes of the main stem and lateral branches.

Determinate types have been described in terms of two main characteristics: those with few nodes and those with many nodes. The few-noded types have between three and seven trifoliate leaves on the main stem before the terminal double raceme; they have been selected for their early maturity and are more suitably grown in cool temperate climates. The many-noded types include those with 15–25 leaves on the main stem. The plant habit is nearly always erect in the determinate type because of the shortness of the internodes, whereas some climbing ability is often present in the indeterminate types because of the long internodes. In these latter types, some have profuse branching and are prostrate and others have reduced branching and are often intercropped with maize in the Andean highlands (Debouck, 1991).

Root development

The root system of *Phaseolus* consists of a fine branching system, which is relatively shallowly developed in open soil conditions. Multiple branching of adventitious roots occurs in non-consolidated soil conditions. Where *Rhizobium* populations occur, most nodule development takes place in clusters around the main root but nodules can develop on any part of the root system and its branches.

Whilst *Phaseolus* will benefit from the symbiotic relationship with soil-borne *Rhizobium* to enable nitrogen fixation to occur, beans grown in non-native areas of the world, particularly in Europe, North America and Australia, where natural populations of specific strains of *Rhizobium* are not present, require additional sources of organic or inorganic nitrogen. Some crops are grown with commercial formulations of *Rhizobium* inoculant applied either to the seed before planting or as granules to the soil, but often this is less efficient and additional nitrogen is frequently necessary

Flower development

Flowers are produced from axillary buds at the nodes of the main stems. These side branches or peduncles can produce several groups of up to three flower buds on bracts produced on one or more racemes. Flowers are typical of Fabaceae in that they are zygomorphic, comprising a standard, two wing petals and the keel, the latter being formed from two fused petals, joined along the anterior edges. Within the flower are the reproductive organs. Flowers are self-pollinating and can vary in colour from white, where the seed coat is also white, to varieties with pink or purple flowers producing brown or speckled or mottled seeds. There are several flowers produced along the axillary bud and these may end in a terminal raceme.

Pod development

Pod setting proceeds acropetally from the base upwards, but flowers may abscise and this can occur on any of the racemes, with up to 50% loss of flowers reported during flowering and pod set. Environmental conditions may affect flower retention but the full range of reasons is not clear. Whilst insect pollinators assist in pollination of *P. vulgaris*, many of the flowers are self-fertile. After pollination, pods develop that botanically are dehiscent fruits, each side of the pod being a carpel. In the natural state, the pods split open along the sutures to eject the seeds.

Pod shape varies between varieties. They may be round or flat in cross-section, or in between. Pod length can vary and so can the width; some varieties picked for the fresh market may be very thin and are known as fine beans. The pod colour is usually green or yellow when young but in some types (borlotti bean, also known as cranberry bean) it may be white with contrasting marbled red markings over the surface. Other types or varieties may be purple or almost black. When mature, beans grown for dry harvest, e.g. navy beans, the pod colour is usually tan. The dehiscent characteristic of the bean pod as means of seed dispersal is due to the presence of lignin along the suture line of the pods; this, along with the fact that maturing seeds are encased within a parchment-like inner wall, is a desirable characteristic for dry-harvested crops but not for crops that are picked as fresh pods for the vegetable market.

Varieties of *P. vulgaris* are either bushy in growth habit or the stems elongate and climb adjacent supporting plants or shrubs. Commercially these have been designated as bush beans or climbing beans. A related species, *Phaseolus coccineus*, has been developed as a mainly climbing bean type and is grown on supports with the green pods picked as a fresh vegetable.

SUMMARY

All legume crops benefit from the symbiotic relationship with *Rhizobium* which ensures a sufficient supply of nitrate through fixation to maintain growth and development of the mature crop. In agricultural systems, this supply of nitrogen is environmentally beneficial by reducing the crop's requirement for inorganic nitrogen application and also benefits a following crop, which is able to utilize the nitrogen residue left in the soil after harvest.

The botany and physiology of peas and *Vicia* beans is similar, though the natural weakness of pea stems is contrasted with the more erect habit of the stiffer-stemmed bean. Pea flowers are self-fertile, whilst pollination by insects vastly improves productivity in *V. faba*. *Phaseolus* beans are very varied in their growth habit and development and this variation provides a wide range of adapted types for both fresh pod and dry bean harvest.

REFERENCES

Anderson, J.S., Rittle, E.J. and Peters, J.C. (2013) Catalytic conversion of nitrogen to ammonia by an iron model complex. *Nature* 501, 84–87.

Baddeley, J.A., Jones, S., Topp, C.F.E., Watson, C.A., Helming, J. and Stoddard, F.L. (2013) Biological nitrogen fixation (BNF) by legume crops in Europe. *Legume Futures Report* 1.5. Available at: www.legumefutures.de (accessed 13 February 2017).

Berry, G.J. and Aitken, Y. (1979) Effect of photoperiod and temperature on flowering peas (*Pisum sativum* L.) *Australian Journal of Plant Physiology* 6, 573–587.

Bond, D.A. and Poulsen, M.H. (1983) Pollination. In: Hebblethwaite, P.D. (ed.) *The Faba Bean (Vicia faba): A Basis for Improvement*. Butterworths, London, pp. 77–101.

Bond, D.A., Lawes, D.A., Hawtin, G.C., Saxena, M.C. and Stephens, J.H. (1985) Faba bean (*Vicia faba* L.). In: Summerfield, R.J. and Roberts, E.H. (eds), *Grain Legume Crops*. Collins, London, pp. 199–265.

Cloutier, J., Laberge, S., Prevost, D. and Antoun, H. (1996) Sequence and mutational analysis of the common *nodBCIJ* region of *Rhizobium* sp. (*Oxytropis arctobia*) strain

N33, a nitrogen-fixing microsymbiont of both arctic and temperate legumes. *Molecular Plant–Microbe Interactions* 9, 523–531.

Debouck, D. (1991) Systematics and morphology. In: van Schoonhoven, A. and Voysest, O. (eds) *Common Beans – Research for Crop Improvement*. CAB International, Wallingford, UK, pp. 55–118.

Delaux, P.-M., Radhakrishnanb, G.V., Jayaramana, D., Cheenab, J., Malbreidl, M., Volkening, J.D. *et al.* (2015) Algal ancestor of land plants was preadapted for symbiosis. *Proceedings of the National Academy of Sciences of the United States of America* 112(43), 13390–13395.

Del Egido, L. (2014) The isolation and molecular characterization of *Rhizobia* from the root nodules of cropped and wild legumes in Scotland in the genera *Vicia*, *Astragalus* and *Oxytropis*. MSc Thesis, University of Edinburgh, September 2014.

Dormer, K.J. (1954) Observations of the symmetry of the shoot in *Vicia faba* and some allied species and on the transmission of some morphogenetic impulses. *Annals of Botany* 18, 55–70.

Emerich, D.W. and Burris, R. (1978) Complementary functioning of the component proteins of nitrogenase from several bacteria. *Journal of Bacteriology* 134, 936–943.

Gallon, J. (1992) Tansley Review No. 44. Reconciling the incompatible: N_2 fixation and O_2. *New Phytologist* 122(4), 571–609.

Howard, J.B. and Rees, D.C. (1996) Structural basis of biological nitrogen fixation. *Chemical Review* 96, 2965–2982.

Iannetta, P.P.M., Begg, G., James, E.K., Smith, B., Davies, C. *et al.* (2013) Sustainable intensification: a pivotal role for legume supported cropped systems. In: *Aspects of Applied Biology* 121, *Rethinking agricultural systems in the UK*, pp. 1–10. Association of Applied Biologists, Warwick, UK.

Iannetta, P.P.M., Karley, A.J., James, E.K., Hawes, C., Begg *et al.* (2015) Benefits of using faba beans for sustainable protein production in Scotland can be calculated. In: *Developing Scotland's Circular Economy*. Ricardo Energy and Environment, Glasgow (Environmental Consultants to the Scottish Government).

James, E. and Crawford, R. (1998) Effect of oxygen availability on nitrogen fixation by two *Lotus* species under flooded conditions. *Journal of Experimental Botany* 49, 599–609.

Kopke, U. and Nemecek, T. (2010) Ecological services of faba bean. *Field Crops Research* 115, 217–233.

Lejeune-Henault, I., Bourion, V., Eteve, G., Cunot, E., Delhaye, K. and Desmyter, C. (1999) Floral initiation in field-grown forage peas is delayed to a greater extent by short photoperiods, than in other types of European varieties. *Euphytica* 109, 201–211.

Minchin, F., James, E. and Becana, M. (2008) Oxygen diffusion, production of reactive oxygen and nitrogen species, and antioxidants in legume nodules. In: Dilworth, M.J., James, E.K., Sprent, J.I. and Newton, W.E. (eds) *Nitrogen-fixing Leguminous Symbioses*. Springer, Dordrecht, the Netherlands.

Nougarède, A. and Rondet, P. (1973) Une modèle original d'organisation de la tige: étude du fonctionnement plastochronique chez *Pisum sativum* L. Var nain hâtif d'Annonay. *Comptes Rendus de L'Academie des Sciences de Paris*, serie D., 277, 997–1000.

Nutman, P.S. (1976) IBP field experiments on nitrogen fixation by nodulated legumes. In: Nutman, P.S. (ed.) *Symbiotic Nitrogen Fixation in Plants*. Cambridge University Press, Cambridge, UK, pp. 211–237.

Oldroyd, G.E.D. and Dixon, R. (2014). Biotechnical solutions to the nitrogen problem. *Current Opinion in Biotechnology* 26, 19–24.

Poulsen, M.H. (1975) Pollination, seed setting, cross fertilization and inbreeding in *Vicia faba* L. *Zeitschrift für Pfl anzenzüchtung* 74, 97–118.

Salunkhe, D.K. and Deshpande, S.S. (1991) *Foods of Plant Origin: Production, Technology and Human Nutrition*. AVI Van Nostrand Reinhold, New York.

Schultze, J. and Drevon, J.J. (2005) P-deficiency increases the O_2 uptake per N_2 reduced in alfalfa. *Journal of Experimental Botany* 56, 1779–1784.

Slattery, J.F., Pearce, D.J. and Slattery, W.J. (2004) Effects of resident rhizobial communities and soil type on the effective nodulation of pulse legumes. *Soil Biology & Biochemistry* 36, 1339–1346.

Smartt, J. (1976) *Tropical Pulses*. Longman, London.

Somerville, D. (2002) *Honeybees in faba bean pollination*. Agnote DAI-128. Reg. No. 166, 26. New South Wales Department of Agriculture, Orange, Australia.

Sprent, J.I. and James, E.K. (2007) Legume evolution: where do nodules and mycorrhizas fit in? *Plant Physiology* 144, 575–581.

Sun, J., Miller, J.B., Granqvist, E., Wiley-Kalil, A., Gobbato, E., Maillet, F. *et al.* (2015) Activation of symbiosis signaling by arbuscular mycorrhizal fungi in legumes and rice. *Plant Cell* 27(3), 823–838.

Svendsen, O.S. and Brødsgaard, C.J. (1997) The importance of bee pollination in two cultivars of field bean (*Vicia faba* L). *SP Rapport-Statens Planteavlsforsøg* 5, 1–8.

Sylvester-Bradley, R. and Cross, R. (1991) Nitrogen residues from peas and beans and the response of the following cereal to applied nitrogen. In: *Aspects of Applied Biology 27, Production and protection of legumes*, pp. 293–298. Association of Applied Biologists, Warwick, UK.

Unkovich, M., Herridge, D., Peoples, M.B., Cadisch, G., Boddey, B. *et al.* (2008) *Measuring Plant-associated Nitrogen Fixation in Agricultural Systems*. ACIAR Monograph 136. Australian Centre for International Agricultural Research, Canberra.

Von Richtofen, J.-S. (2006) What do European farmers think about grain legumes? *Grain Legumes* 45, 14–15.

Ward, R. and Palmer, S. (2013) A study of soil nitrogen supplies available to winter cereals following pulses compared to winter cereals following oilseed rape. *Book of Abstracts of First Legume Society Conference 2013: A Legume Odyssey*, Novi Sad, Serbia, 9–11 May 2013, p. 273.

White, J., Prell, J., James, E.K. and Poole, P. (2007) Nutrient sharing between symbionts. *Plant Physiology* 144, 604–614.

Wright, A.T. (1990) Yield effect of pulses on subsequent cereal crops in the Northern Prairies. *Canadian Journal of Plant Science* 70, 4, 1023–1032.

3

PEA AND BEAN BREEDING

BACKGROUND TO THE CURRENT TYPES

There is a significant number of similarities in the genetic, physiological and adaptational characteristics of leguminous food crop species that allows them to be considered together as well as genus by genus. The most significant historical work on peas (*Pisum sativum*) was carried out by Mendel (1866). Although his work was overlooked by most applied botanists until its rediscovery at about the same time by Correns (1900), de Vries and Tshermack in Germany and William Bateson in Cambridge (Bateson, 1901; Druery and Bateson, 1901), it remains fundamental to genetic understanding of all studied plant species and animals. Peas are a largely self-pollinated and hence inbreeding species, as is the common bean species *Phaseolus vulgaris* (but notably not *Phaseolus multiflorus* syn. *P. coccineus*). Wild landraces (now regarded as locally adapted ecotypes) of such largely inbreeding species comprise mixtures mainly of homozygous plants and of heterozygotes from crosses that have occurred naturally as a result of insect pollination, which is facilitated by the form of the flowers and availability of nectar. Dry beans were studied by W.L. Johannsen in Denmark, who established the different effects on genotypes when influenced by environmental factors, i.e. phenotype (Pierson, 2012). The work also related to trait expression in barley and provided information relating to inbreeders where 'pure line' based populations are required by seedsmen and merchants requiring stable varieties for marketing. *Vicia faba*, however, is largely outbreeding, with individual flowers on any plant being easily pollinated by pollen from other flowers on the same or other plants in the population. In outbreeders, self-pollination can be enforced by human manipulation but the resulting plants from the seeds produced are often weaker or otherwise different from typical plants of the parental population. Line breeding in these species does not produce satisfactory new varieties but 'inbreeding depression' is common, while forms of 'mass selection' allow useful named populations to be produced, grown and marketed. However, significant improvement can be made by a strategy of 'synthetic' variety production whereby a variety is produced by crossing, in all combinations, a number of inbred lines that combine well with each other. Once synthesized, a synthetic is maintained by open pollination in isolation. This strategy has been emulated in several bean-producing countries.

A second set of important similarities concerns the diseases of these three crop species and the breeding strategies relevant to their management and control. These diseases may arise from soil-borne infections, infective aerial spore dispersal, bacteria dispersed to the surrounding plants by rain or splash dispersal or seed-borne fungal or bacterial pathogens. As a result, disease resistance breeding has figured as a major part of overall plant breeding efforts. Locating, documenting and sometimes effectively using sources of resistance to disease has played a substantial role in large germplasm collections during the past 70 years but was initiated from two directions: the Vavilov Institute's collections for the former Soviet Union, and those of the United States Plant Introduction Services catalogued by the Plant Introduction (PI) numbers. There is also the curating of these collections, such as *Pisum* at the John Innes Institute in Norwich, UK, and particularly the handling at Cambridge University of a sub-collection of the PI set used in Uganda through an intended 'Aid' initiative (Leakey, 1970).

PEA BREEDING

Within *P. sativum* there is much genetic variation among kinds grown in Europe. A related species, *Pisum abyssinicum*, is apparently not separated by any genetic barriers but produces finely speckled and very starchy pea seeds. These are old African mountain kinds and because of ancient use of varieties based on local ecotypes, there was much material of interest to Mendel. His distinctions between tall and dwarf plants, round and wrinkled seeds and yellow and green colour are still the basis of descriptors of all peas produced commercially.

Although Mendel's work with peas was instrumental in defining the basis of genetics, there were many recorded instances of pea varieties being available by natural selection from natural variations as long ago as the 16th century. But it was not until the mid-18th century that the first known crosses of selected lines were made to produce varieties. This provided a wide range of pea varieties for Mendel to work with. He concentrated on variables such as the pod colour and characteristics, the presence of string or parchment in the pod and the shape and colour of the seeds, together with selection and breeding for those with long internodes producing tall varieties and those with short internodes and the reduced internode length that produced a fasciated group of flowers at the top of the plant.

With recent improvements in the study of pea genomics, such techniques as marker-assisted selection (MAS), which identifies a particular location within a plant's DNA sequence to identify specific markers, tend to underpin all traits. These include disease resistance, plant architecture, seed quality and other characteristics. Such techniques will improve the efficiency with which breeders can select plants with a desirable combination of genes throughout the variety development process.

Afila peas

Between 1968 and 1984, work at the John Innes Centre in the UK explored the agronomic potential of a series of genetic traits controlling foliage characteristics, most notably in a number of independent spontaneous mutations from Argentina, Russia and Finland where examples of peas in which the leaflets had been replaced by tendrils (*afila* gene) created significant new commercial innovations in both vining and combining peas. This characteristic had the potential to assist with mutual plant support, as the tendrils were able to intertwine and the risk of crop lodging could be reduced. It was also thought that this would reduce disease risk in wet conditions and provide more uniformity of ripening. The *afila* (*af*) mutation in pea has been widely used since the early 1970s following the suggestion that the additional tendrils might assist with standing ability of the combined crop (Snoad, 1974). The fully 'leafless' pea was one of the initial ideotypes for a dried pea breeding programme at the John Innes Institute and later was widely taken up by private breeders. The first 'leafless' variety was released in the UK in 1978 as the cultivar 'Filby'. The tendrils comprise replaced leaves and stipules carried on the stems of the leafless ideotype. This 'leafless' ideotype was able to produce sufficient photosynthate not to limit the development of the sink, i.e. seeds, at normal agronomic plant populations. However, the ideotype was found to be limited when grown at low planting densities and so there was a limitation to the total biomass of plants and hence the potential yield (Hedley and Ambrose, 1981).

Semi-leafless peas

The solution was to retain the *afila* allele but revert to the use of wild-type stipules. The increased photosynthetic area overcame the problem of the fully leafless form and the 'semi-leafless' model has been adopted by most pea-growing countries of the world for the past 30 years as parental material, to develop a large number of commercially successful varieties for fresh market, home and garden, freezing and canning and dry harvest combining peas.

Stem stiffness

Aerial stem stiffness is of great agronomic importance for commercial production of peas as the trait enables the crop to remain upright (particularly during periods of adverse weather and through to harvest), reduces the risk of disease and greatly assists the mechanical harvesting operation. Pea stems remain inherently weak at the base and breeders are continuing to improve on this now that the genetic variation for this characteristic has been identified (Zhang *et al.*, 2006).

Disease resistance

Disease resistance has also been a principal goal for plant breeding. Peas are susceptible to several major fungal pathogens that are commonly found in many of the larger growing areas and in particular where peas have been cultivated for many years. Fungi have a propensity to evolve quickly and develop into strains that overcome some of the resistance characteristics. In the case of pea powdery mildew (*Erysiphe pisi*) resistance has been found to be due to three genes, which have been introduced into many modern varieties. So far the original single resistance gene identified in 1975 is still holding up well and either completely or partially resistant varieties are available and used widely.

Pea bacterial blight (*Pseudomonas syringae* pv. *pisi*) is known to exist in several race forms. Most European-bred varieties have complete resistance to race 2 but not to race 5.

Wilt diseases caused by soil-borne root-infecting fungi such as *Fusarium oxysporum* can occur as a result of the build-up of the pathogen in the soil following pea cultivation over a long period. *F. oxysporum* occurs in several races, the most common in Europe being races 1 and 2, whilst in USA several other races (4, 5 and 6) are commonly found. Resistance to these races has been developed that also appears to be fairly robust (Porter *et al.*, 2014). With other pathogens, varieties have been selected from germplasm showing various degrees of tolerance. In the case of pea downy mildew (*Peronospora viciae*) a number of races and pathotypes have been identified (Stegmark, 1994) and it has not been possible to introduce completely resistant varieties, though good levels of tolerance to the predominant races in certain regions are now being achieved. Resistance to other diseases such as pea enation mosaic virus and pea seed-borne mosaic virus has recently been developed.

Very light and sandy soils are unlikely to retain sufficient moisture in dry seasons and the yield of peas is severely compromised by drought stress. Whilst the *afila* characteristic is thought to be beneficial in reducing the effects of drought stress, even between *afila* varieties there are indications of varietal difference of tolerance to drought (E. Uber and A. Biddle, 2011, unpublished). Often drought-stressed crops become more susceptible to diseases affecting the root and vascular system, particularly *Fusarium* wilt (*F. oxysporum* pv. *pisi*). Currently, a multi-partner research project in the EU is aimed at developing crops that have increased resistance to drought and *Fusarium* stress (ABSTRESS, 2013).

Colour

The colour of peas, especially for freezing, must be bright and uniformly green. In some seasons, where excessive vegetation has been produced, the presence of a proportion of pale or 'blonde' peas can occur. Some varieties are more susceptible to this than others, though in general *afila* types are less prone to such problems. The loss or retention of green colour by vining and marrowfat pea seeds is an economically significant quality parameter. Work is advanced in identifying markers for green cotyledon colour that may be more resistant to chlorophyll loss or bleaching. Lines with stable green colour have been identified and used to develop recombinant inbred lines with associated maps to define genetic loci involved in seed colour stability (Domoney *et al.*, 2009).

Flavour

Flavour is usually mild and characteristic of pea and varieties should not exhibit any off-flavours or bitterness. Sweetness has been defined through marker technology (Domoney, 2011). Biochemical changes in seed composition during maturation are being defined and quantified. These not only allow specific stages of development of the pea to be associated with compounds and groups of compounds, but also point to significant differences among cultivars in the relative amounts of compounds that are likely to influence quality. The same is true of the content of a group of compounds known as saponins, linked frequently with bitterness characteristics of foods. Differences in saponin content between developmental stages and cultivars are likely to be relevant to quality, and this information is being used by breeders to produce lines of peas with these desirable characteristics.

Genetic manipulation

Most new pea varieties are obtained using the traditional method of crossing and selection. There have been some developments in genetic manipulation to produce lines with a specific insect resistance from *Phaseolus* by a group in Australia but this material was never commercialized and no commercial genetically modified pea varieties are in use currently. Novel variation is being produced by non-genetically modified means through the use of induced mutation programmes and some developments arising out of this method have been commercialized, such as in the production of super-sweet pea lines and herbicide resistance.

Vining pea varieties

Vining peas can be grouped into sweet and starchy types. Sweet varieties have compound starch grains and seed which in the dry state is deeply wrinkled. They are used for canning and freezing in the UK and other European and Scandinavian countries, the USA, Canada, Australia and New Zealand. Starchy peas have simple starch grains, the dry seed is smooth and spherical and they are mainly grown for canning in southern Europe. The processing properties of the two types are quite different. In addition to being less sweet, smooth-seeded peas have a mealy texture and the canned product may also contain herbs and spices.

Most vining or fresh market pea varieties used commercially are white-flowered. The breeding of white-flowered varieties is made easier by the fact that peas are self-pollinating and are therefore not at risk of cross-pollination from insects, etc.

Vining peas, in practice, are grouped according to the time taken between sowing and maturity. A difference of 16 days is found between the earliest and latest wrinkled-seeded varieties, which are classed as 'first early' and 'main-crop' groups, respectively. Between these two extremes are the 'second early' and 'early maincrop' groups.

Varieties are then described according to their agronomic characteristics. Harvest date is referred to as relative maturity, or days earlier or days later in harvest than a standard named variety. Extreme earliness may be a character-istic that is useful for production, though yield is often sacrificed for earliness. Extreme lateness may only be useful for the fresh market sector, because pro-cessors cannot extend their processing season in the factory as it may clash with the optimum time for harvesting other vegetables.

Pea size can vary from season to season but generally the seed size of a specific variety will tend to be within its mean range. In Europe, seed size is defined by diameter, with large peas having a diameter >10.2 mm, medium size >8.75–10.2 mm, small size >7.5–8.7 mm and very small (also known as petits pois) <7.5 mm.

Virtually all commercial varieties are of relatively determinate habit, i.e. flowering and haulm usually terminate, allowing relatively even maturation. Many varieties in commercial use in Europe are semi-leafless, though a num-ber of conventionally leaved varieties are also grown. The semi-leafless charac-teristic is not always favoured by growers in drier climates, such as the USA, as there is a perception that these types are more prone to drought. Earliness and haulm length are often linked, as earliness can be achieved by flowering at a lower node. Consequently it is sometimes the case that early varieties can be too short and the later ones too long. Haulm length is also influenced by soil type, district and season but it is not such a problem with the use of complete pea harvesters (viners). Very tall varieties are often more prone to lodging, despite having the *afila* characteristic. Lodged crops are more susceptible to fungal decay when wet conditions are likely. Lodged crops are also difficult to harvest by hand for the fresh market.

Disease susceptibility is an important characteristic of vining pea varieties as it can seriously affect the quality of the pods and vined peas. Resistance to powdery mildew (*Erysiphe pisi*), race 1 *Fusarium* wilt (*Fusarium oxysporum* f. sp. *pisi*) and relative tolerance to pea downy mildew (*Peronospora viciae*) are examples of some resistance traits available.

The varieties grown for large-scale commercial freezing or canning are usually specified by the processor, often in consultation with the growers. Ideally varieties within a harvesting programme should all give peas of similar size and colour, unless a proportion of other types is required for a specific purpose, so that if two varieties have to be processed together, the resulting product is acceptable. Most processing factories deal with peas over a 6–8-week period. To achieve the necessary length of season, several varieties are required, each falling into the appropriate maturity groups. Control of sowing time is also necessary to provide the required harvest period and this area is examined later.

Most pea breeders worldwide participate in variety trials in various areas of the pea-producing countries. These trials are often carried out by the breeders themselves or they allow their varieties to be evaluated and compared with others in independently run trials funded by growers, processors, public bodies or universities. Each trial provides data on field performance and in some cases assessments are made on the processed quality of the peas after harvest. An example in the UK is variety trials carried out by the Processors and Growers Research Organisation (PGRO, 2014).

There are then a large number of varieties available for use as vining or fresh market peas. The varietal characteristics of particular importance include their time of maturity for harvest (which is a relatively stable characteristic), the pea size profile (again a very stable character), haulm length and disease resistance. Environmental factors have an influence on crop growth and, in particular, haulm length and yield. Because of this, the value of data derived from variety trials needs to be considered in relation to the geographical area or soil type of the intended crop. Nevertheless, such data are useful as they are often accompanied by data from a well established variety for comparison.

Table 3.1 lists some of the characteristics of a selection of modern standard-sized varieties in commerce in the UK by 2015 (PGRO, 2015). Note that these include both conventionally leaved and semi-leaved varieties. These data

Table 3.1. Varieties of vining peas available in the UK (from PGRO, 2015).

Season	Variety (SL=semi-leafless)	Days earlier (−) or later (+) than the standard	Haulm length (cm)	% Peas large (>10.2 mm)	%Peas medium (>8.75–10.2 mm)	Resistance to downy mildew	Resistance to powdery mildew
First earlies	Pizarro^{SL}	−1	54	31	45	Slightly susceptible	Susceptible
	Avola	0	57	44	42	Susceptible	Susceptible
	Twinkle	0	45	34	50	Susceptible	Susceptible
Second earlies	Jaguar	+5	53	30	49	Susceptible	Resistant
	Chinook	+6	44	19	58	Moderately susceptible	Susceptible
	Novella^{SL}	+8	43	30	53	Moderately susceptible	Resistant
	Biktop^{SL}	+9	43	29	57	Good field resistance	Susceptible
Maincrop	Oasis	+11	57	35	50	Moderately susceptible	Susceptible
	Ambassador	+12	68	41	46	Moderately susceptible	Resistant
	Kiros	+12	54	20	52	Moderately susceptible	Susceptible
Late	Columbus	+16	59	21	50	Slightly susceptible	Resistant

are derived from UK trials and illustrate varieties with various maturity and seed size categories.

As can be seen from the examples in Table 3.1, many of the varieties for vining are conventionally leaved types. Also the majority are susceptible to downy mildew and this is a concern in temperate areas where peas have been grown over a long period. Control at present relies on the use of a fungicidal seed treatment. Powdery mildew infects crops at a late stage in the season and is only of relevance to those varieties planted late. The main features of petits pois varieties is that most of the produce falls into the small (>7.5–8.7 mm) and very small (<7.5 mm) seed size categories. Within most of the commercially available varieties grown in Europe, the maturity dates are very similar and there is less scope to have a fully integrated programme of sowing and harvesting dates (see Chapter 4).

Fresh market pea varieties

In the majority of production systems, the peas are hand harvested and in some cases the peas are picked over several times before the crop has reached maturity. This is particularly the case with varieties grown as mangetout types where the peas are grown in beds and may be supported on wires or trellises. Multiple harvesting may also be the preferred method in much smaller production systems. With the exception of mangetout or sugar snap types, there has been very little breeding effort for fresh market varieties and there is a reliance on either older established varieties or specific varieties chosen from the wider range of vining peas that are currently available. There are specific criteria for fresh peas that focus on pod characteristics. Length, colour, size and number of peas per pod are the main specifics and these may vary depending on the requirements of the retailer. In addition, haulm length, disease resistance and time of harvest are important considerations. Early-maturing varieties tend to be short in haulm length and also lower yielding than later-maturing types. Resistance to powdery mildew is an advantage for late-sown crops as the disease is favoured by warm dry day temperatures and high humidity at night, conditions that are more typical in late summer. Examples of early, maincrop and late varieties used for hand picking for the fresh market include Early Onward, Feltham First, Elvas, Onward, Kelvedon Wonder, Hurst Greenshaft and Ambassador.

Pea pods lacking more than minimal wall fibre and little or no string formation when the pods are young contribute to the mangetout types, which generally fall into two categories: those harvested as flat pods before the peas develop in size are known as snow peas (Fig. 3.1); and those harvested as plump pods, where the seeds have developed inside the pods, are known as sugar snaps. In both types, the preference is for stringless pods where there is little or no parchment inside the pods and an absence of the membranous string

Fig. 3.1. Mangetout peas.

running along the top of the pod from the base to the tip (McGee and Baggett, 1992). Pod colour also varies and there are a number of mangetout types with green, yellow or purple pods.

In the early days of pea selection in France, a tall-growing snap-podded type was discovered and commercialized in the USA to produce a range of snap peas such as Sugar Rae, Sugar Bon and Sugar Mel while yellow-podded varieties such as Golden Sweet and purple-podded Shiraz are now available. Flat-podded varieties include Oregon Sugar Pod and Snowbird.

In the 1940s it was found that there was an easily extractable arabinoxylan, gum tragacanth, a form of soluble dietary fibre that is present in pod walls of snap peas and which was thought to give a smooth texture in the mouth when eating sugar snap peas. This led to plant breeding research which tried to develop shorter-haulmed snap peas with a high density of pods but whose fresh snap pods at the appropriate harvesting time could be mechanically harvested. The variety Cygnet was registered but was never effectively developed.

Combining pea varieties

Combining peas are harvested in the dry state to be used either for human consumption or in the compounding of animal, fish or pet foods. In the UK, the combine-harvested dried pea varieties mainly comprise white-, blue- and green-seeded marrowfat peas. Most are now of the semi-leafless ideotype. In many countries, varieties are mostly white flowered but a small number of coloured-flowered varieties are grown for speciality markets. In the tropics, however, many coloured-flowered varieties are produced for the human consumption market locally and are also grown in the USA and Canada for export to the Indian subcontinent and Africa. Coloured-flowered varieties produce light brown skin colour, often flecked with darker pigmentation. Usually these varieties are round and smooth skinned.

Combining peas are generally classified according to flower colour and seed type. Seed size is dependent on growing conditions but a small number of small-seeded varieties are grown. Of the white-flowered varieties, the marrowfats are blue-green, large, dimpled and dented. Varieties in this group are important for human consumption, being used for dry packet sale, export for use as pea snack food or canning as large 'processed peas'. They are suited to a wide range of soil types but are relatively late maturing and less suitable for northern temperate areas. They are also susceptible to pre-harvest bleaching and require care at harvest time to maintain quality. Most marrowfats are grown in the UK (Fig. 3.2).

White peas have a white/yellow seed coat and are usually round. Although mostly used for animal feed, some are used in canning, as pea flour or in prepared meals. White-flowered varieties are usually the highest-yielding peas

Fig. 3.2. Marrowfat peas.

and are suitable for a range of soil types. They are widely grown in Europe, the USA and Canada (Fig. 3.3).

Blue peas are so called because of the blue-green colour of the seed and green cotyledons. Usually the seed is round and smooth skinned. They are used for canning and also for micronizing in the pet food industry. Most varieties have a short straw length, stand very well in the field up to harvest time and have good disease resistance. Yields are generally high but blue peas have the potential of early maturity.

Examples of combining pea varieties grown in UK and Europe are shown in Table 3.2.

Table 3.2 demonstrates the main agronomic characteristics that are now used extensively for combining pea production. Straw strength is of major importance to allow successful mechanical harvesting and disease resistance to pea downy mildew reduces the need for chemical seed treatment.

The seeds of coloured-flowered varieties may be dimpled and dented or smooth and round. The produce is used in speciality markets or for human consumption. Mainly coloured-flowered varieties are grown in tropical countries, where the peas are used as a staple source of protein in dhal or falafel meals. Older varieties are conventionally leaved types and are indeterminate; they are used as whole-crop silage (Fig. 3.4).

Fig. 3.3. White peas.

Table 3.2. Varieties of combining peas available in the UK (from PGRO, 2015).

	White peas			Large blue peas			Maple peas		Marrowfat peas		
	Salamanca	Gregor	Mascara	Prophet	Crackerjack	Daytona	Mantara	Rose	Sakura	Neon	Genki
Earliness	5	6	6	5	5	6	5	6	5	4	4
Standing ability at harvest*	7	6	5	6	5	6	6	6	5	4	6
Resistance to downy mildew*	6	6	7	7	5	7	7	7	5	6	5
Thousand-seed weight (g)	270	317	290	303	305	285	242	261	396	351	360
Protein content %	23.1	24.2	22.3	21.7	23.1	23.1	22.9	25.9	23.4	22.2	23.8

* A high rating indicates that the variety shows the character to a high degree (0–9)

Fig. 3.4. Coloured-flowered peas.

Autumn-sown (winter) peas

Most varieties in northern countries are spring sown, but in countries where rainfall is in short supply during summer months or temperatures are excessively high, then some varieties have been bred for being suitable for autumn planting. However, winter-sown peas can be susceptible to freezing injury and disease. Successful varieties have a long vegetative period to withstand cold conditions and ideally should be grown in areas where temperatures remain low to prevent excessive vegetative growth (Eteve, 1985).

Experience of winter peas in continental Europe has revealed their susceptibility to disease in mild winter conditions, especially *Mycosphaerella pinodes*, which causes severe leaf stem and pod spot. In addition, the seed-borne pathogen causing pea bacterial blight can be a problem when late frosts damage the leaf tissue at the beginning of flowering. An approach has been made to breed varieties with a delayed flowering date to avoid the risk of frost damage (Eteve *et al.*, 1999). In Europe the variety Lucy, developed by INRA France, has had some success and experience in growing winter peas and durum wheat as a mixture has also had some success in France.

In the USA autumn-sown coloured-flowered peas (dun or Austrian winter) are sometimes grown, especially where water supply is limited in the summer months. Autumn-sown peas are also often grown as forage crops, or as green manure in these areas.

VICIA FABA

Background to modern varieties

There is little history of a scientific approach to faba bean breeding until quite recent times. This may be because it is an out-pollinated species where pure line approaches to breeding are not relevant, nor are there easily worked out Mendelian character markers. Hence varieties grown and marketed were not often precisely defined or clearly recognizable from one another. It is believed that *V. faba* may have been derived from two or three ancestors of the *Vicia* family, *V. narbonensis*, *V. galilaea* and *V. hyaeniscamus*, but there is very little evidence to show that it is related to these three (Ladizinski, 1975). It has also not been possible to cross *V. faba* with any of these species and so far this has limited the variability available for breeding varieties of *V. faba*, especially as there is good resistance to some of the common diseases of beans in *V. narbonensis*.

V. faba has remained a distinct species in its domesticated form, though some of the smaller-seeded types have been known as *V. faba* subsp. *minor* and the larger types as subsp. *major* or *equina*. There is, however, no evidence that there are any barriers whatsoever to gene flow between seed size classes and therefore botanically these should not be classed as subspecies, as the arbitrarily defined seed size ranges are more or less just market classes. The sources of variation that do exist in *V. faba* are through its wide domesticated gene pool, which has been helped to some extent by the popularity of beans, especially the large-seeded beans as garden vegetables and as a field crop harvested dry. There has been some mutation variation induced by chemical or radiation treatment (Sjodin, 1971) but, generally, current varieties of *V. faba* have been generated through normal selective breeding methods.

The recent development of di-haploids through anther and pollen culture has proved applicable to *V. faba* and at least one variety is currently being marketed.

Although there is some autofertility of flowers on the plant, *V. faba* is partially allomagous and can outcross by cross-pollination effected by insects, mainly bumblebees (*Bombus* spp.). This has several important consequences for bean breeding that should be noted. The development and maintenance of pure lines requires the complete exclusion of pollinators, which constitutes a large overhead on the breeding programme and more or less makes it impossible to release pure inbred varieties. The use of male sterility to guarantee maximum hybrid vigour has been tentatively explored but a commercially viable hybrid production system has never been achieved. The compromise is to exploit the natural tendency for between one-third and two-thirds of seeds to result from outcrossing by creating so-called 'synthetic' varieties in which a number of founding individuals that share some common characteristics, but are by no means homologous, are allowed to inter-cross freely, giving a degree of hybridity that may decline over generations of multiplication in the absence of optimal pollinator populations (Obiadalla-Ali *et al.*, 2015).

Beans have been able to adapt to environmental conditions and as they were more widely grown in areas north and south of the Mediterranean, varieties arose based on exploitation of their adaptations rather than specific breeding methods. Adaptations include winter hardiness, reduced pod shatter (dehiscence), seed size and seed colour variation. Varieties of beans that have been more recently developed include those with winter hardiness so that they can be sown in the autumn, those that are spring sown, more determinate growth of the stem to reduce crop lodging, medium seed size and a light brown to green testa colour. The green-seeded character, as a result of persistent chlorophyll, has been identified as a single recessive gene-determined characteristic and has been used in vegetable broad beans in China and elsewhere.

In spite of its popularity since its domestication, the faba bean area decreased very much during the last century as a consequence of the introduction of mechanical labour, in that faba bean was a main ingredient for feeding oxen, horses and mules. There was also very little effort put into plant breeding to produce new varieties that were better adapted to modern agriculture. This meant that beans were constrained to have low yield and poor yield stability, particularly during periods of drought or excessive moisture, indeterminate growth leading to excessive stem growth and susceptibility to lodging, relatively low disease resistance and the presence of anti-nutritional factors, thereby reducing their suitability as a feed for many types of livestock. However, as the interest in the crop began to increase following the desire to produce more sustainable sources of vegetable protein and the value of grain legume crops in arable crop rotations and in less developed agricultural systems, some recent significant advances have been achieved in both breeding and agronomy.

Modern cultivars have a more determinate habit (i.e. ceasing continued vegetative growth when flowering and pod setting are advancing), low anti-nutritional factors and increased resistance to the more common diseases, including ascochyta (*Ascochyta fabae*), rust (*Uromyces viciae-faba*) and downy mildew (*Peronospora viciae*), and resistance to broomrape (*Orobanche crenata*). Advances in the study of the genomics are producing a gene map and marker-assisted selection is also in progress as well as work on synteny with other legumes such as pea and *Medicago truncatula*.

The full genetic potential of *V. faba* is still being studied, including leaf architecture, pod wall characteristics, stem determinacy, factors affecting flower abortion and pod set, and seed quality aspects such as protein content and a reduction of anti-nutritional factors, including tannins (Stoddart, 1986).

Because of the indeterminate habit, stems can produce excessive growth at the expense of reproductive nodes. Competition for light and pollinators results in poor flower retention and the production of undersized pods at the top of the stems, making the crop slow to mature and difficult to harvest. The development of a type with terminal inflorescence (ti type) was introduced to breeders. This type had a stiff stem with the inflorescence at the top of the stem. However, this was somewhat extreme and the type was less able to adapt to variable climatic conditions during growth and is not widely used. The identification of markers for determinate types is in progress (Alvila *et al.*, 2007).

The most important fungal foliar diseases are chocolate spot (*Botrytis fabae*), *Ascochyta fabae* and *Uromyces viciae-fabae*. In some areas the soil-borne root-infecting fungi *Rhizoctonia solani* and *Fusarium* species occur, especially in soils with a long history of faba bean production. *Botrytis* is seen in a wide range of growing conditions; it is often a very serious threat, but there is very little information on a source of resistance, though several less susceptible bean genotypes are known (e.g. ICARDA lines originating from Ecuador). For *Ascochyta* and *Uromyces*, specific resistances are known and molecular markers have been developed. Downy mildew causes further, less well studied foliar diseases in *V. faba* but, as with the case of peas, there are several races of the pathogen and no complete resistance to mildew is known. Breeders rely on selection of less susceptible cultivars for commercialization. Viruses can be a frequent problem and virus diseases may occur as epidemic and become serious. Pea enation mosaic, bean yellow mosaic virus, bean leaf roll virus, broad bean true mosaic virus and broad bean stain virus are frequently found. The effects of aphid-transmitted viruses can be reduced by efficient crop protection methods but the seed-borne viruses can be more of a problem. The most important pests are *Aphis fabae* (the black aphid) and *Acyrthosiphon pisum* (the pea aphid). In addition to direct feeding damage, virus transmission occurs where aphids introduce infection early in the growing season and no useful resistance is known.

Faba bean may be infested by the stem nematode *Ditylenchus gigas*, which is widespread and is common especially in cases where nematode-infested seed is used. Recent work has identified potential breeding material that shows resistance to the nematode and development is in progress to evaluate the potential of this material to provide a robust level of resistance. Broomrape (*Orobanche crenata*) is a parasitic plant, damaging pulses and other crops in the Mediterranean Basin and Nile Valley. Partially resistant genotypes are available; the trait shows a quantitative genetic variation and recently three quantitative trait loci (QTLs) for resistance have been identified (Torres *et al.*, 2006). There are few known sources of resistance against root rot. Varieties with low tannin in the seed coat have been introduced but as these varieties are usually white flowered, the difficulties of multiplying homozygous stocks make commercial introduction very costly.

The degree of cross-pollination by bees and other flying insects is about 50%, with a large genotypic and environmental component of variation and with marked heterosis where the heterozygous types show on average less outcrossing; and in an open field situation, seed production suffers from contamination with cross-pollination, unless spatial isolation and pure lines are used.

Despite the limitations for breeding the perfect bean, breeders have made significant progress in introducing a steady stream of new varieties for commercial production. In Europe, significant advances with yield improvement have been made and the newly introduced varieties Fanfare and Vertigo have clearly shown this improvement and are currently being grown on a significant scale.

Varieties of faba beans

Field beans
Faba beans grown for dry harvest (field beans) are usually known as either winter beans where they are sown in the autumn and are more frost hardy, or spring beans, which are usually sown as early as possible in the spring to achieve their full yield potential.

The main use of the seed is as a source of vegetable protein for animal feed and aquaculture, though dried faba beans also form an important part of the human diet in Africa and the Middle East. The seed quality for the human consumption market is important in that the beans must be plump rather than flat, with an even light-brown skin colour and a pale-coloured (known as white) hilum. Because of the interest from European and Australian producers to supply these markets, most varieties of faba beans have been selected for these characteristics (Fig. 3.5).

For animal feed, there are no specific requirements for colour or shape and the beans may also have a black hilum, though the presence of severely decayed broken seeds or high levels of insect damage may downgrade the value of the crop. The presence of anti-nutritional factors such as tannins have been thought to be detrimental to a high inclusion rate of beans in compound feeds, but recent work has shown that the presence of tannins from beans in pig feeding rations did not result in negative growth rate or digestibility in fattening pigs.

In the UK, a small market exists for the racing-pigeon feed market, which requires a small round seed. Currently one variety, Maris Bead, is grown for this purpose and this variety has been in commercial use since 1964. Because of the very small market for these beans (known as tic beans), no breeding effort has been expended in recent years.

There have been recent developments of faba bean varieties with enhanced nutritional qualities, including tannin-free varieties (for example Gloria), and with low vicine and convicine. These two glycosides are associated with poor digestion rates in animals and the disorder known as favism in humans where individuals harbour a deficiency in the activity of their glucose-6-phosphate dehydrogenase.

These low-vicine types have been registered in France by INRA (*Institut National de la Recherche Agronomique*) and are known as Fevita types. Some of these may be preferred by specialist animal feed manufacturers, though in the UK and Canada there seems to be no commercial advantage in these types so far.

Varieties of both spring and winter beans are tested in field variety trials in the UK and in several European countries, Australia and Canada. Current

Fig. 3.5. White hilum faba beans.

varieties are listed in the UK as a Recommended List and updated annually (Table 3.3).

Broad beans

Within *V. faba*, there are two main types of bean based on their uses that are described here: the beans that are harvested at the immature green stage where the seeds are removed from the pod and consumed as a vegetable; and the type that is harvested when fully mature to obtain the dry seed, used for a number of purposes in human or animal food. The vegetable bean, known as the broad bean, is grown either for the fresh market or for large-scale commercial

Table 3.3. Varieties of field beans currently recommended in the UK (from PGRO, 2015).

(a) PGRO Recommended List of Spring Beans

	Vertigo	Fanfare	Boxer	Fury	Fuego	Pyramid	Babylon	Maris Bead (tic)
Yield as % of control	106	104	102	101	99	97	95	87
Flower colour	c	c	c	c	c	c	c	c
Earliness of ripening *	7	7	7	7	7	8	8	6
Shortness of straw*	6	6	6	7	6	6	7	5
Resistance to downy mildew*	5	5	4	6	5	5	7	7
Thousand-seed weight (g)	558	518	542	500	542	542	521	374
Protein	27.7	28.7	27.7	28.1	28.0	27.4	27.3	29.6

(b) PGRO Recommended List of Winter Beans

	Wizard	Honey	Arthur	Clipper
Yield as % of control	97	93	103	102
Flower colour	c	c	c	C
Hilum colour	white	white	black	black
Earliness of ripening*	8	9	8	6
Shortness of straw*	8	9	7	6
Standing at harvest*	7	9	5	5
Thousand-seed weight (g)	704	702	664	687
Protein %	27.2	26.1	26.1	25.4

*A high figure indicates that the variety shows the character to a high degree.

freezing or canning. Varieties have been produced for the appearance, flavour, texture and colour of the immature seed and this can range from a pale green to a dark green, either very large in size or slightly flattened in shape to small and almost round beans. Crop height also varies between varieties. Very short dwarf-type varieties, such as The Sutton, have very few vegetative nodes and begin their reproductive stage on nodes that are still very low to the ground. Often these varieties are earlier maturing than the taller types. Pods are formed in tight inflorescences around the stem. Some varieties are more frost tolerant and can be sown in the autumn to provide an early harvest date. Seed set can be a problem with autumn-sown varieties as insect activity at flowering may be low in cool periods and this can result in irregular seed set within the pod. Aquadulce types are often autumn sown for the fresh market.

Two basic types of broad beans are grown: those with white flowers (Fig. 3.6), which are suitable for canning; and those with coloured flowers, which are suitable for freezing, though most varieties grown for freezing are white flowered.

Beans for the fresh market can be either white- or coloured-flowered types. Coloured flowers indicate the presence of leuco-anthocyanins which break down during cooking to coloured polyphenols, giving pink or brown beans in cloudy brine when canned. The freezing market requires a bean with even seed size and colour, while the fresh market also requires a pod width and length of defined ranges of dimensions, a minimum number of beans per pod and a seed to pod ratio of around 35%.

Fig. 3.6. White-flowered broad beans.

PHASEOLUS BEANS

Dried beans

Dried beans are the most important class of beans throughout the world and are so widespread that they are often generally referred to as a group that contains beans of different classifications and species. The main types considered in this book are *Phaseolus vulgaris*, or common beans, which include white or navy, pinto, kidney, black, pink, romano, yellow and several other types, but there are a number of other species, including mung beans (*Vigna radiata*) and black matpe or urad beans (*Vigna mungo*), which are a different genus and therefore are not included.

Their origins have been traced from countries ranging from northern Mexico to northern Argentina and the cultivars have arisen from many years of domestication of wild types with varying seed characteristics. Types with the smallest seed size have been associated with wild populations whilst those with larger seed have been further domesticated and cultivated. Much work has been carried out on the genetic background of the species using specific traits, including the seed protein phaseolin, to attempt to identify the origins of specific *Phaseolus* cultivars (Gepts *et al.*, 1986).

Many studies were carried out by Kaplan (1965a,b) on ancient plant remains of four *Phaseolus* species in Middle Americas. Those studies suggested that *P. vulgaris* may have originated in Mexico about 700–400 years ago. It now seems that beans originate from two main sources: the Central American types, which were mostly small seeded; and Andean types, which were generally large seeded. Both types were disseminated throughout the main growing areas of Central and South America, Africa, Europe and the USA.

Bean breeding focused on the optimization of some specific characteristics, mainly morphological and physiological. The most obvious morphological difference is the climbing growth habit of the wild plant. Climbing habit provides an advantage in wild populations of mixed plants, as a climbing stem is able to compete for light amongst woody plants and trees, which provide support for a climbing bean. Once in contact with a support, the climbing bean twines around and produces branches that grow at right angles to the force of gravity. These climbing types also produce profuse branches, long internodes and high node numbers. Most climbing forms are also very indeterminate.

Cultivation led to a different set of selection pressures as a greater degree of determinacy to allow an easier harvest. This also led to a shortening of the internodes to produce a more compact plant and one that had sufficient stem strength to stand alone, with a loss of twining without the need for additional support.

A system of descriptors of the different plant types was drawn up with data from the International Centre for Tropical Agriculture (CIAT, 1980) and the PI collection in the USA based on seven types (IBPGR, 1982) and this has subsequently been modified by Leakey (1988).

These descriptors are summarized in Table 3.4.

Cultivation in northern latitudes, such as Europe, has led to a different set of selection pressures from those in the tropics and subtropics. This has been possible through day neutrality allowing a greater degree of determinacy and earliness of maturity. Day neutrality allowed shortening of the internodes to produce a compact stiff-strawed plant type.

The growing environment has also influenced the selection criteria in other climate regimes in different countries and growth pattern ideotypes favour (or

Table 3.4. A suggested description of *Phaseolus* plant types (from Leakey,1988).

Description	Cultivars or cultivation group
Indeterminate climber, long internodes, inflorescences mostly at top of plant	Primitive Andean type (*P. aborogineus*) ancestors
Climbing with well distributed pods; limited branching, not day-length sensitive	European and American climbing beans
Indeterminate short internodes; branches long, little or no twisting	Many strong-stemmed bush beans, North American Pintos and early European Swiss beans
Indeterminate bush; branches short, main stem growth and terminal guide little developed; no twisting	Refugee, common in Middle East germplasm
Indeterminate bush; vigorous, contorted stem; short internodes; lateral branches equal growth	Black-seeded American cultivars. Early navy bean types
Erect, indeterminate bush; strong untwisted main stem and branches; little development of a distinct guide	San Fernando, typical of selected Middle American cultivars
Determinate growth; reproductive buds terminating main stem and branches; vigorous and spreading bush	Many Spanish and Portuguese types
Spreading determinate bush; twining in main stem and branches	Several Bush Blue Lake types
Determinate bush; weak branches, little terminal bud	Masterpiece, Diacol, Nima, etc.
Determinate bush; vigorous multi-podded; erect upright lateral branches; short lower nodes	Chimbolo
Determinate bush; open habit; few well developed branches.	Michigan pea bean
Determinate bush; contracted lower internodes; densely branched; concentrated zone of internal flowering	Most modern bush varieties of French and Dutch breeding, suitable for mechanical harvesting

otherwise) differences such as low rainfall, high temperatures and altitude. In northern latitudes, day neutrality is important when the growing season is short. Photoperiodism is the effect of day or night length on growth and inflorescence and in beans the inheritance of the response is controlled by at least two major loci expressing dominance or recessiveness, or both (Smartt, 1988).

Leaf size can have an influence on susceptibility to drought stress and larger leaf areas are advantageous in areas of more diffuse light, at high altitude or in oceanic climate areas of the world.

Pod dehiscence is a characteristic that is disadvantageous and cultivars that have a low incidence of seed shedding have been preferred. Pod walls that contain a high fibre content are more prone to splitting and selection of low-fibre pods has reduced the dehiscence characteristic while increasing the palatability of the pods when harvested green and consumed as a vegetable.

Seed size and colour are also important characteristics and selection of specific criteria have resulted in the wide range of varieties used today. Although the wild types often had a range of seed colours varying from black to brown, including mottled testa, domestication and subsequent selection led to cultivars with larger seeds than the wild types and a wide range of seed colour and patterns, including reds, cream and white.

Commercial dry bean classes in North American literature are usually described in terms of their colour and size. These market class names are now recognized internationally in trade. Table 3.5 shows descriptions of dry beans.

Red- and garnet red-seeded dry beans

A recessive red allele produces the garnet red of dark red and also light red kidney beans (Smith, 1939). Other red beans such as the red Mexicans and others carry the dominant red allele. Many of these reds lose their colour during cooking but recently a small garnet red bean, variety Stop, with the same size and shape as a Mexican small red and bred with the same gene background as a dark red kidney has been developed that retains its colour during cooking (Leakey, 1999).

White-seeded beans

During the 1960s four navy bean cultivars from an X-ray-induced breeding programme at Michigan State University were introduced and assessed as a potential canning crop in the UK (Kelly, 2014). There was some commercial interest and one of the varieties, Seafarer, was re-registered as Purley King and trialled for production (Scarisbrick et al., 1976; Evans and Davis, 1978). Although at the time it was not commercially successful, development is continuing with further varieties.

Great Northern beans

Attempts at introducing Great Northern beans to France was seriously affected by seed-borne infections of bacterial common blight (*Xanthomonas phaseoli*).

Table 3.5. Descriptions of dry beans (from Goodwin, 2003).

Colour	Names	No. of seeds per 100 g
White	Great Northern Large White	280–330
Black	Black Turtle Black Bean Preto	500–550
Pink	Pink	330–400
Reddish brown	Dark Red Kidney	150–200
Red-brown mottled on white	Pinto	260–300
White	White Kidney Alubia	150–200
Pale red	Light Red kidney	170–220
Light brown	Dutch Brown	210–300
Red	Small Red Red Mexican	275–330
White	Navy White Pea Bean	450–525
Red mottled on white	Cranberry Borlotti	145–225

Since then, disease-resistant germplasm of Great Northern beans has been developed and subsequent breeding work was carried out to develop an early maturing, determinate type with some success, in a hybridization programme with Leakey's Horsehead (Leakey, 1999).

Brown-, yellow- and green-seeded dry beans

The colour of Brown Swedish and Brown Dutch beans is due to glucosides of the flavonoid quercetin. The yellow-coloured beans characteristic of the seasonally very dry Pacific coastal areas of South America contain no quercetin, but instead the flavonoid kaempferol and its monoglucoside. The yellow beans are known as Canarios, Mantecas and Mayacoba and are reputed for their digestibility, as they contain no tannins. The variety Prim was a result of a wide cross and has seen some commercial production (Leakey and Harbach, 1975).

Green-coloured seed, due to persistent chlorophyll, was originally selected as French *flageolets verts* and many green-seeded flageolets have been bred in France, some varieties incorporating anthracnose (*Colletotrichum lindemuthianum*) resistance (Dean, 1968).

The nutritional quality of the seed is also being studied and improvements in protein content and quality have been identified as potential goals. The reduction in anti-nutritional factors is also desirable, as the presence of trypsin inhibitors and lectins has an influence on digestibility. These are heat labile and beans require a period of full re-hydration followed by thorough cooking to denature these substances before consumption. Lectins persist in undercooked or slow-cooked red kidney beans.

Improvements in seed yield by increasing seed size, decreasing pod shatter, standing ability and earliness of maturity are all major breeding objectives.

Early maturity is an important characteristic, particularly where beans are grown in the more northern temperate areas of the world when frosts can cause severe damage to crops as they are reaching maturity. Cold tolerance is needed for vigorous early germination (soil temperatures not less than 12°C are required); and cool temperatures during the early stages of growth result in short plants and hence pods that develop close to the ground or on the soil surface, making recovery during mechanical harvesting difficult, with a high degree of pod shatter and seed loss. Heat and drought tolerance in later stages of growth are often desirable in hot summers.

As well as these morphological and physiological characteristics, breeders have sought to develop cultivars with increased resistance to a wide range of pests, diseases and disorders. Major pathogens include bean rust (*Uromyces appendiculatus*), white mould (*Sclerotinia sclerotiorum*), some viruses and bacterial blight (*Pseudomonas* spp.). Breeding for resistance to pathogens, particularly in the tropics, is very difficult as there is a large number of pathogens present; each one may be capable of seed transmission and almost all may be present in several race forms or strains. Wild *Phaseolus* types are currently being screened for their resistance to pests and diseases and some lines have been identified for future development.

In France and the USA, the problem of disease avoidance is largely overcome by producing seed in areas that are free from disease due to the warm dry weather conditions that persist throughout the vegetative growth season. Seed production in Africa follows a similar practice.

Green beans (dwarf French, snap beans)

Fibreless pods or 'snap' beans are widely grown and used as a frozen product. Pod size and shape vary greatly between cultivars. For the processed vegetable market, the pod size determines the end product (Table 3.6).

SUMMARY

Whilst all three species have many similarities, the range of types and varieties that have been used and are currently grown have developed from programmes of selection of suitable plant types that have been grown in different environments and geography. Breeding techniques are still based on traditional crossing techniques, though this is difficult in *Vicia* where cross-pollination by

Table 3.6. Pod dimensions for green French and Romano beans (data from PGRO, 2013).

Category	Dimensions	Common name
Extra fine	6–8 mm	Kenya beans
Fine	8–9 mm	Bobby beans
Medium	9–10 mm	Filet beans
Large-podded	>10 mm	French and some fleshy Romano types
Flat-podded	Often >15 mm	Italian flat beans

insects can affect the production of pure line breeding material, but there has been little or no commercial investment in genetic manipulation to produce suitable varieties. However, the identification of marker technology is helping to make significant advances, especially in pest and disease resistance.

REFERENCES

ABSTRESS (2013) Improving the resistance of legume crops to combined abiotic and biotic stress. EU Seventh Framework Programme Project number FP7-KBBE-2011-5-289562. Food and Environment Research Agency (Fera), York, UK.

Alvila, C.M., Atienza, S.G., Moreno, M.T. and Torres, A.M. (2007) Development of a new diagnostic marker for growth habit selection in faba bean (*Vicia faba*) breeding. *Theoretical and Applied Genetics* 115(8), 1075–1082.

Bateson, W. (1901) Experiments in plant hybridization by Gregor Mendel. *Journal of the Royal Horticultural Society* 26(1), 1–32.

CIAT (1980) *Description of Growth Habits of* Phaseolus vulgaris L. Annual Report, Beans International Centre for Tropical Agriculture 1980.

Correns, C. (1900) G. Mendel's Regel über das Verhalten der Nachkommenschaft der Rassenbastarde. *Berichte der deutschen botanischen Gesellschaft* 18, 158–168. (English translation: Piternick, L.K. (1950) G. Mendel's law concerning the behavior of progeny of varietal hybrids. *Genetics* 35(5), 33–41; see also: Stern, C. and Sherwood, E.R. (1966) *The Origin of Genetics: A Mendel Source Book*. W.H. Freeman, San Francisco, California, pp. 119–132.

Dean, L.L. (1968) Progress with persistent-green colour and green seed coat in snap beans (*Phaseolus vulgaris*) for commercial processing. *Experimental Horticulture* 3, 177–178.

Domoney, C. (2011) Understanding pea seed quality. *The Vegetable Magazine*, winter 2011, p. 5. Processors and Growers Research Organisation, Peterborough, UK.

Domoney, C., Chinoy, C., Laugier, F., Charlton, A. and Clement, A. (2009) The challenge of manipulating seed quality traits in pea for multiple end uses. *Proceedings of North American Pulse Improvement Association (NAPIA)*, October 28–30, 2009, Fort Collins, Colorado.

Druery, C.T. and Bateson, W. (1901) Experiments in plant hybridization. *Journal of the Royal Horticultural Society* 26, 1–32.

Eteve, G. (1985) Breeding for cold tolerance and winter hardiness in pea. In: Hebblethwaite, P.D., Heath, M.C. and Dawkins, T.D. (eds) *The Pea Crop – A Basis for Improvement*. Butterworths, London, pp. 131–138.

Eteve, G., Lejeune Henault, I., Bourion, V., Cunot, E., Delhaye, K. and Desmyter, C. (1999) Floral initiation in field grown forage peas is delayed to a greater extent by short photoperiods than in other types of European varieties. *Euphytica* 109, 201–211.

Evans, A.M. and Davis, J.H.C. (1978) Breeding *Phaseolus* beans as grain legumes for Britain. *Applied Biology* 3, 1–42.

Gepts, P., Osborne, T.C., Rashka, K. and Bliss, F.A. (1986) Phaseolin-protein variability in wild forms and landraces of the common bean (*Phaseolus vulgaris*): evidence for multiple centres of domestication. *Journal of Economic Botany* 40(4), 451–468.

Goodwin, M. (2003) Crop profile for dry beans. *Pulse Canada*, February 2003.

Hedley, C.L. and Ambrose, M.J. (1981) Designing 'leafless' plants for improving the dried pea crop. *Advances in Agronomy* 34, 225–277.

IBPGR (1982) Phaseolus vulgaris *Descriptors*. International Plant Genetic Resources Unit, Rome.

Kaplan, L. (1965a) Archeology and domestication in American *Phaseolus*. *Economic Botany* 19, 358–368.

Kaplan, L. (1965b) Domestication of *Phaseolus* spp. In: Smartt, J. (ed.) *Tropical Pulses*. Longman Higher Education, London.

Kelly, J.D. (2014) *One Hundred Years of Bean Breeding at Michigan State University: a Chronology*. Michigan State University Occasional Publication. Michigan State University, East Lansing, Michigan.

Ladizinski, G. (1975) On the origin of the broad bean *Vicia faba* L. *Israel Journal of Botany* 24, 80–88.

Leakey, C.L.A. (1970) The improvement of beans (*Phaseolus vulgaris*) in East Africa. In: Leakey, C.L.A. (ed.) *Crop Improvement in East Africa*. CAB International, Slough, UK.

Leakey, C.L.A. (1988) Genotypic and phenotypic markers in common bean. In: Gepts, P. (ed.) *Genetic Resources of Phaseolus beans*. Springer, Dordrecht, the Netherlands, pp. 245–327.

Leakey, C.L.A. (1999) Progress in developing *Phaseolus* beans for Britain. In: *Aspects of Applied Biology* 56, *Protection and Production of Combinable Break Crops*, pp. 195–202. Association of Applied Biologists, Warwick, UK.

Leakey, C.L.A. and Harbach, C. (1975) Beans, fibre, health and gas. In: *Agri-food Quality – an Interdisciplinary Approach*. Royal Society of Chemistry, Cambridge, UK, pp. 175–180.

McGee, R.J. and Baggett, J.R. (1992) Inheritance of stringless pods in *Pisum sativum* L. *Journal of the American Society of Horticultural Science*, 117(4), 628–632.

Mendel, J.G. (1866) Versuche über Pflanzenhybriden Verhandlungen des naturforschenden Vereines. In: Brünn, Bd. IV für das Jahr, 1865 Abhandlungen 3-47. English translation, see: Druery, C.T and Bateson W (1901) Experiments in plant hybridization. *Journal of the Royal Horticultural Society* 26, 1–32.

Obiadalla-Ali, H.A., Mohamed, N.E.M. and Khaled, A.G.A. (2015) Inbreeding, outbreeding and RAPD markers studies of faba bean (*Vicia faba* L.) crop. *Journal of Advanced Research* 6, 859–868.

PGRO (2013) *Green bean varieties – 2013 update*. Processors and Growers Research Organisation, Peterborough, UK.

PGRO (2014) *PGRO Pulse Agronomy Guide 2014*. Processors and Growers Research Organisation, Peterborough, UK.

PGRO (2015) *PGRO Vining Pea Growers Guide 2015*. Processors and Growers Research Organisation, Peterborough, UK.

Pierson, B.R.E. (2012) Wilhelm Johansen's genotype-phenotype distinction. *Embryo Project Encyclopaedia*. ISSN: 1940-5030. Available at: http://embryo.asu.edu/ handle/10776/4203 (accessed 13 February 2017).

Porter, L.D., Kraft, J.M. and Grünwald, N.J. (2014) Release of pea germplasm with *Fusarium* resistance combined with desirable yield and anti-lodging traits. *Journal of Plant Registrations* 8(2), 191–194.

Scarisbrick, D.H., Carr, M.K.V. and Wilkes, J.M. (1976) The effect of varying plant population density on seed yield of navy beans (*Phaseolus vulgaris*) in south-east England. *Journal of Agricultural Science* 86(1), 65–76.

Sjodin, J. (1971) Induced morphological variation in *Vicia faba* L. *Hereditas* 67, 155–179.

Smartt, J. (1988) Morphological, physiological and biochemical changes in *Phaseolus* beans under domestication. In: Gepts, P. (ed.) *Genetic Resources of Phaseolus Beans*. Kluwer, Dordrecht, the Netherlands, pp. 143–161.

Smith, F.L. (1939) A genetic analysis of red seedcoat color in *Phaseolus vulgaris*. *Hilgardia* 12, 553–621.

Snoad, B. (1974) A preliminary assessment of 'leafless peas'. *Euphytica* 23, 257–265.

Stegmark, R. (1994) Downy mildew on peas (*Peronospora viciae* f sp *pisi*). *Agronomie* 14(10), 641–647.

Stoddart, F.L. (1986) Pollination, fertilization and seed development in inbred lines and F1 hybrids of spring faba beans. *Plant Breeding* 97, 210–221.

Torres, A.M., Roman, B., Avila, C.M., Satovic, Z., Rubiales, D. *et al.* (2006) Faba bean breeding for resistance against biotic stresses: towards application of marker technology. *Euphytica* 147, 67–80.

Zhang, C.Z., Tar'an, B., Warkentin, T., Tullu, A., Bett, K.E., Vandenberg, B. and Somers, D.J. (2006) Selection for lodging resistance in early generations of field pea by molecular markers. *Crop Science* 46(1) 321–329.

AGRONOMY OF PEAS AND BEANS

Peas and beans have specific requirements for successful crop establishment, growth and yield of high-quality produce. Whilst there are many similarities in agronomy, particularly between peas and *Vicia faba*, the requirements for *Phaseolus* beans are not too dissimilar. The successful management of crops is dependent on a number of factors, including the use of high-quality seed, providing a suitable location and soil conditions for sowing and the supply of adequate nutrients and moisture to maintain growth. In addition, crops must be protected from weed competition and pests and diseases; these subjects are discussed in Chapters 5 and 6.

CROP ROTATION

There are a number of traditional reasons put forward for the necessity of crop rotation when including peas or beans. Weed control may be improved by the use of spring-sown crops and there is the value of residual nitrogen (see Chapter 2) to improve fertility of the soil for the following crop. Large-seeded legumes fit in well with a rotation with cereals and they can generally be grown using the same machinery and stored with existing equipment. The vegetable crops are more demanding on machinery, labour and harvesting equipment but nevertheless their value as a break crop and general soil improver is still the same.

The establishment of soil-borne pests and diseases depends to a large extent upon the closeness of cropping with peas or beans or other host crops. Nematode pests (see Chapter 6) can affect peas or faba beans and are soil borne or, in the case of stem and bulb nematodes, can also be introduced to the field through infested seed. The fungal pathogens causing downy mildew and those involved in the soil-borne root-infecting disease complex such as *Fusarium solani* f. sp. *pisi*, *Didymella pinodella* (syn. *Phoma medicaginis* var. *pinodella*) and *Aphanomyces euteiches* are likely to increase in intensity with prolonged cropping.

Both peas and beans can be found within the same arable rotation and because there are commonalities between the pest or pathogen and the legume host crop, peas and beans should be classed as the same crop for the purposes of rotation. The longer the period in the absence of host crops that the rotation can allow, the less likely it is that soil-borne populations of pathogens and pests will build up

Currently, in the UK and Europe, the minimum period between large-seeded legumes is 4 years, during which neither peas nor beans are grown. In practice, in the UK experience has shown that a longer break is beneficial to reduce the risk of losses by the root disease complex.

Vining peas

Large-scale commercial production of green peas for freezing and canning requires a greater degree of planning and management throughout the growing and harvesting operations. Peas generally prefer soil types of light to medium texture that are free draining and are not susceptible to compaction, waterlogging or drought. In view of the capital investment involved, the limited throughput of harvesting machinery and processing equipment and the fact that vining peas are of optimal quality for a very brief period, the use of a full range of different varieties, soils and conditions to prolong the harvesting season and a planned sowing programme to include the range of maturity available are necessary for maximum returns.

Fields are selected for specific sowings of vining peas in relation to their soil type and aspect, as these have an influence on the rate of development and maturation of the crop. An early start to sowing is preferred in the UK and temperate regions, though in more southern areas of Europe some autumn sowing takes place in order to avoid the high temperatures of the following season and subsequent effect on moisture availability and quality. On medium- and heavier-textured soils the moisture in the seedbed is usually adequate for the crop through to flowering. Up to that time, rainfall has little influence on yield unless it is excessive and adversely affects soil structure.

Combining peas

Combining peas are almost always grown as a spring crop in eastern and northern Europe, the USA and Australasia, though autumn planting of specialist winter-hardy varieties is made in areas that have a warmer and drier summer climate. Early sowing can make a considerable contribution towards high yield, but cold wet soil conditions early in the spring are detrimental and therefore sowing should be made later, when soil conditions have improved. The quality of combining peas, especially those grown for human consumption, is usually superior to that for those sown later, as weather conditions are generally better during harvesting and there is therefore less staining of the seed caused by secondary saprophytic fungi, which can colonize pods before senescence.

Faba beans

Beans can tolerate slightly heavier soil types than peas, which makes them more versatile for cropping in a range of arable situations. The seed is well able to withstand prolonged cool conditions before germinating; however, the bean is susceptible to drought stress and early establishment of the spring crop allows the development of an extensive root system, which helps to withstand dry conditions.

Autumn-sown varieties, known as winter beans, can be planted in late autumn on heavier more moisture-retentive soils to allow establishment of the root system before winter. A period of low temperatures over winter encourages the production of multiple branches of shoots, which are compact in early spring until they elongate as temperatures increase. Winter beans are planted at a lower density than the spring varieties to allow for the increased number of stems, all of which are productive.

Overwintering of autumn-sown beans effectively makes them more susceptible to foliar diseases caused by fungal pathogens such as *Botrytis fabae* and *Ascochyta fabae* but modern varieties are much more tolerant of these fungi, which can also be managed by chemical treatment during the growing season.

Phaseolus beans

The diverse areas of production and growing systems for *Phaseolus* beans provide a wide range of agronomic requirements and management practices, whether the beans are grown as large-scale green vegetables for processing, hand picking for the fresh market, as dry beans for direct sale or for processing as canned beans. The choice of soil type is important for the rapid establishment of what is a relatively short-season crop and this restricts the large-scale production of *Phaseolus* beans to specific geographical areas where efficient mechanization is available.

SEED ESTABLISHMENT AND SOWING

Vining and fresh market peas

The establishment and development of some pests and diseases of peas are influenced by the extent of close cropping with peas and with other host crops.

The longer the rotation, the less likely it is that there will be a build-up of soil-borne pests or diseases. Research in the 1970s indicated that peas should not be grown in a field more frequently than once in 5 years (Biddle, 1983). However, since that time it has become more general practice, particularly in the UK and Europe, to extend that rotation to 6 or even 7 years. Foot rot diseases caused by a complex of soil-borne fungi have been shown to have a serious effect on yield. The complex commonly found includes *Fusarium solani, Didymella pinodella* (syn. *Phoma medicaginis* var. *pinodella*) and *Aphanomyces euteiches*. All of these, once established in soil, take a long period of time to decline and once high levels are detected, then pea cropping may be uneconomical for 10 years or more. Similarly, pests such as pea cyst nematode (*Heterodera gottingiana*) can persist as viable cysts in soils for around 15 years.

The pea seed consists of the seed coat (testa), two cotyledons and the embryo axis. The testa protects the cotyledons and embryo axis. At the point of detachment from the seed stalk is a scar known as the hilum and the small micropyle that allows the ingress of water during the early stages of imbibition. Pea seed must have the ability to germinate and produce a strong, vigorous seedling, often under cool, wet conditions. It is the ability of a seed to perform in this way that is described as having a good seed vigour. Vining peas often have to be sown in very early spring, when establishment of seedlings can take up to 6 weeks, though very little of the germination process occurs at temperatures below 4°C. Seeds that exhibit low seed vigour often become infected by soil-borne fungi such as *Pythium* spp. which results in the seedling failing to emerge. Wrinkle-seeded vining peas are much more susceptible to this failure than round-seeded or marrowfat varieties of combining peas.

Seed tests have been developed to provide an indication of seed vigour and the electrical conductivity test is almost universally accepted as a valid test method (ISTA, 2006). The test is dependent on the fact that, on being soaked in water, carbohydrates and inorganic salts are released from the seed. The greater amount of leachates lost, the lower is the degree of seed vigour and subsequently the more likely it is that the seed will fail to emerge in adverse conditions. The test measures the electrolytes in the water after the peas have been soaked for 24 h. The interpretation of test results provides an indication of seed performance under a range of conditions.

In the UK, a set of pea seed vigour grades was evolved and is now extensively used by the seed industry. The grades are based on the units of electrical conductivity (microsiemens) measured in 250 ml of water after 50 pea seeds have been allowed to soak for 24 h. The grade is derived from the reading divided by the dry seed weight and expressed as microsiemens (µS) per gram. Table 4.1 provides the vigour grades and their interpretation for use (PGRO, 2014).

There are several factors that can cause low vigour in vining peas, the most important being the integrity of the seed coat. Microscopic cracking of the testa during the harvest or seed-processing operations can result in rapid

Table 4.1. Vigour grades for wrinkle-seeded vining and fresh market peas.

Reading	Vigour grade	Use
>24.0 µS/g	High vigour	Seed suitable for sowing early
24.1–29.0 µS/g	Medium vigour	Some seedbed losses may occur in adverse conditions but can be used for later sowings without problem
29.1–43.0 µS/g	Low vigour	Not suitable for early sowing and may fail in cold wet conditions
<43.1 µS/g	Very low vigour	Not suitable for sowing

imbibition, which damages the cells of the cotyledons, thereby releasing excessive amounts of salts and sugars. These damaged areas become rapidly infected by soil-borne fungal pathogens, especially *Pythium* spp. (Matthews, 1971). Commercially, seeds are often treated with a fungicide to help to protect them from such losses, but a seed of low vigour is likely to fail despite such protection.

Seed-borne diseases, especially *Ascochyta pisi* and *Mycosphaerella pinodes*, can also adversely affect field emergence of pea seedlings. Again the use of healthy seed is of high importance, though some reduction of low levels of seed-borne fungal pathogens can be achieved by an application of systemic fungicide seed treatments.

In large-scale vining pea production, seed is sown mechanically using cultivator drills into previously ploughed land. The depth of sowing depends on the time of sowing but generally the depth is around 5 cm after rolling, or slightly deeper later in the season when the soil is drying. For fresh market crops, sowing is again largely made using seed drills as for vining peas, but in small-scale production hand sowing in split furrows or similar may be carried out.

Combining peas

Seed is usually round or dimpled. Compared with wrinkle-seeded vining peas, the seed is not as susceptible to seed-coat injury and is not as susceptible to pre-emergence mortality through low seed vigour. Although attempts have been made to identify seed lots with a propensity for low seed vigour, the amounts of leachates lost during imbibition are not as great and therefore the electrical conductivity test has not been found to have relevance to combining pea seed (Bladon and Biddle, 1991). The seed is much more robust and can emerge successfully from a wide range of soil conditions. In practice, a fungicide seed treatment is commercially applied but in many instances this may not be a necessity unless control of seed-borne *Ascochyta* spp. is required.

Sowing is usually done on land that has been ploughed the previous autumn to allow winter frosts to weather the soil. Spring cultivation is then carried out immediately prior to sowing, to minimize compaction. As an alternative, direct drilling into minimally cultivated stubbles is carried out in some situations, though currently this is not widely practised. Peas are drilled at a depth of 4–5 cm, with the earlier sowings a little more shallow and the later sowings a little deeper where the soil moisture is low.

Faba and broad beans

The seed of *V. faba* is generally more robust than other pulse species, the testa is thicker and the seed size is generally larger. Coloured-flowered beans have tannins within the testa and these have anti-fungal activity, which can be of value to the seed when sown in cool wet soil conditions where infection by soil-borne *Pythium* spp. is favoured; it may also have an effect of reducing the susceptibility to *Fusarium* spp. such as *F. culmorum* and *F. solani*, but not all tannin-free beans are susceptible (Villalobos and Jellis, 1990). Although in some instances there is a relationship between field emergence of seedlings and seed vigour as expressed by an electrical conductivity test, in general seed appears to perform adequately if the initial seed germination capacity and health are satisfactory. However, as a large seed, it is susceptible to mechanical injury, particularly during harvesting, handling and artificial drying, and all of these factors can adversely affect the germination of seed. Seed damage by the larvae of bruchid beetles (*Bruchus rufimanus*), which leave holes in the seed, may also affect germination or subject damaged seed cotyledons to infection after planting.

An important pest of *Vicia* beans is the stem and bulb nematode *Ditylenchus gigas*, which is transported in the seed and, once planted, the resulting seedling and its neighbours become infested by nematodes released from the original seed. Freedom from this pest is recommended for all *V. faba* seed.

Crop rotation is important, as soil-borne root-infecting fungi can build up if beans are cropped frequently in the field. Also populations of pea cyst nematode (*Heterodera gottingiana*) and broomrape (*Orobanche crenata*) and others, once introduced, can survive in soil for periods of time.

Bean seed is usually planted early in the season on a wide range of soil types. Good drainage is important, especially for autumn sown beans although these are usually sown into heavier soils, which are less prone to drought. Spring-sown beans are usually planted as early in the season as possible to ensure an adequate root system can develop before the onset of summer where moisture stress may be an issue.

Winter beans can be planted at a slighter greater depth than spring crops to avoid predation by corvids or game birds. However, as with all peas and beans, they are sensitive to compacted soil layers. There are a number of methods of planting field beans, including a conventional seed drill, or broadcasting seed on the surface and ploughing down, or by direct drilling in zero-tillage or minimum-till operations. It is usual to plant on previously ploughed land but in western Europe more planting is carried out using minimum cultivation and non-inversion tillage. In this way, the seeding depth is kept more constant, allowing an even seed distribution and emergence as most seed is drilled to a depth of around 8 cm.

In developing countries or where small areas of crop are grown, broadcasting of the seed is a normal practice, e.g. in Ethiopia and Lebanon (Lahoud *et al.*, 1979) and to a smaller extent in other countries. Furrow planting, where seed is placed in split furrows and then harrowed, or where the seed is individually dropped into small holes made with a small dibbing spade, is carried out in others. In the main, sowing is mechanized using seed drills with variable depth control and large-sized coulters to prevent seed lodging.

Phaseolus beans

The seed is very fragile and very susceptible to mechanical damage to the seed coat or to the cotyledons, where damage to the dehydrated embryonic plumule can occur as a fracture at the hypocotyl. Seed with such damage emerges with the absence of a growing point, resulting in the condition known as baldhead or snake head. The testa is also very thin and easily damaged. Germination is hypogeal, where the radicle expands pushing the cotyledons above the soil surface before the expansion of the cotyledon leaves. Germination is temperature dependent and very little of the process begins at soil temperatures below 10°C.

Phaseolus beans require conditions to allow a rapid germination and emergence of the cotyledons. In some cases, seed that is planted early in cool conditions can suffer from poor emergence and several attempts have been made to develop a laboratory-based seed vigour test. As with peas, the electrical conductivity test has been shown to provide a relationship between high electrolyte leachates and low field emergence, but setting a grade for a test has been difficult due to large differences in seed size. There appears to be a requirement for a separate set of results to interpret the vigour of large-seeded types compared with small-seeded types. Another vigour test used as an advisory tool in the UK relies on the use of tetrazolium chloride solution whereby the abaxial surfaces of the cotyledons are allowed to soak in a solution after removing the testa and the number of cotyledons with areas of unstained or non-respiring tissue are noted. The relationship between high numbers of seeds with unstained areas and poor field emergence has been shown to apply in the UK and is used to identify those seed-lots with poor potential for early sowing.

The beans are very sensitive to compaction and especially soil capping, which can occur on light soil types with poor soil structure, and to soil consolidation, which will impair root development. Beans are grown in a diversity of cropping systems, including large-scale commercial production of dried beans or beans for the processed vegetable market but also on a smaller scale with less investment in mechanized cultivation, sowing and harvesting.

In Europe, the USA and other developed areas, large-scale planting operations depend on precision drills, which are able to plant beans at a uniform depth with uniform spacing both between seeds and between rows. They may be planted in wide beds of around 2 m across or with even distribution of rows across the field. Such planting requires a well-formed seedbed with adequate moisture to allow even and rapid germination and emergence.

In other areas, planting is determined by temperature and the onset of rainy seasons and altitude (Woolley *et al.*, 1991). In tropical and semi-tropical areas, beans may be intercropped with maize or other crops and as such have different approaches to establishment. In these areas, both bush beans and climbing beans may be grown either solely or intercropped.

PLANT POPULATION AND SOWING RATE

Vining and fresh market peas

Seed comprises the largest single growing cost of the crop and much work has been in progress to maximize the efficiency of production by determining the optimum plant population for maximum return. As plant density is raised, yield increases rapidly at first and then more slowly until a point is reached where no further yield increase takes place. As populations are increased beyond this point, yield gradually falls and the value of any yield increase gained by raising plant density through increasing the seed rate must be balanced against the cost of the extra seed (King, 1967). The development and introduction of *afila* varieties may also affect the optimum plant population that is required, but little experimental work has been carried out with sufficiently robust results to determine the specific optimum requirements for different varieties or types. In the UK and Europe, the most efficient plant population is around 100 plants/m^2 (PGRO, 2015). Although most vining peas are planted in rows 10–15 cm apart, there is an advantage in using a more precise spacing to allow even growth of individual plants. Work carried out with semi-precision drills in the UK showed that there was a significant evenness of maturity when harvested at the freezing stage, probably by reducing the propensity of the stems to branch, where a more precise placement of seed was made; in addition, precision sowing brought the harvest date forward by around 2 days with no compromise in yield (Smith, 2007).

Large-scale commercial production of vining peas is carried out with the use of efficient herbicides and without the restrictions on row widths, which had been previously determined by the need for inter-row cultivations. The advantages of closer spaced rows in providing a more even distribution of plants across the field are the improved competitiveness of peas with weeds and the even maturation of the crop. However, the availability of old-established herbicides is becoming restricted, with an increased awareness of the possible negative effects on the environment by the use of persistent active ingredients in soil-applied herbicide. Currently, row widths of around 20 cm or less are commonly used. As the availability of selective herbicides becomes more limited, recent use of mechanical weed control using spring-tined cultivators operating at a shallow depth or self-guided weed cultivators that operate between the rows may have a considerable impact on production.

With fresh market peas, the different growing systems can define the methods of establishment, the row width, plant spacing and the required population. For the unsupported fresh market pea crop, the crop may be drilled as for vining peas using narrow row widths and a target plant population of 100 plants/m^2. Some systems require multi-harvesting and access to the crop by hand labour. Peas may be drilled in beds typically 1 m wide to allow access by foot. In early-planted crops or autumn-planted peas, the rows may be protected by fleece and narrow beds utilized. For supported crops, such as mangetout types, the rows are wide enough to gain access between the trellises or wire supports. Plant spacing remains at around 2.5 cm.

Combining peas

The larger-sized seed of combining peas and the length of growing season from planting to full maturity has necessitated a different requirement for the optimum population of the dry-harvested crop. In addition, the range of combining pea types has determined different populations for each. The response to increasing plant density may be due to the different growth characteristic of the varieties rather than the seed size; however, changes in the plant spacing can affect growth habit and, with a decreased population, pea stems are encouraged to develop side branches which may produce enough pods to make a positive contribution to yield, which is why the optimum densities for combining peas are lower than for vining peas. In the UK, the optimum plant density for the larger seed size in marrowfat types is 65 plants/m^2 whilst for the round-seeded white or large blue types the optimum is 70 plants/m^2 (PGRO, 2015). As most commercial combining peas are harvested mechanically, there is no requirement for planting in wide rows unless inter-row weeding methods are used.

Faba beans

Faba beans harvested dry comprise autumn-sown and spring-sown varieties. The growth habit differs depending on the planting density, with the need to produce a small seedling of autumn beans at a stage of development to be protected from the winter and a rapidly growing spring bean plant to enable the establishment of a deep root system to withstand possible summer drought. For the autumn-sown crop, a definition of the optimum plant population is complicated because the establishment of plants that survive to the following spring can be more variable than in spring-sown crops. Also disease pressure from fungal pathogens such as chocolate spot (*Botrytis fabae*) can alter the yield response to plant density. Recent work in the UK has shown the effect of early sowing and dense planting on the incidence of chocolate spot. With a reduced plant population, chocolate spot is much less likely to affect growth early in the spring. Similarly the time of sowing influences the sowing rate of beans, particularly in autumn sowings. Beans planted in early autumn are more susceptible to chocolate spot at higher densities (Maguire, 2013). In autumn-sown beans, low plant densities allow more weed competition, increased branching on the earlier sown crops and pods that are produced lower on the stem, increasing the loss of yield due to poor pod collection at harvest. At high densities, there is an increased disease risk, an increase in the risk of lodging on moisture-retentive soils and the potential of yield reduction due to poor pollination and seed set (Armitage, 2009).

Successful crops are usually judged on their establishment and the number of surviving plants in the spring. Many older varieties of winter beans were indeterminate and susceptible to lodging at densities in excess of 18 plants/m^2. With the adoption of more compact varieties, there is evidence that higher populations are more economically viable and populations of up to 25 plants/m^2 are common in the UK. As with all optimum population curves, an increase in plant density increases the seed cost and therefore the optimum economic seed rate takes the seed cost into account. Figure 4.1 illustrates the relationship of increasing seed rate and yield and shows the yield response and the economic response curves. The results are taken from recent work in the UK with the winter bean variety Wizard (Armitage, 2009).

In spring beans, the variability is less than in winter beans but time of sowing in spring and establishment methods also influence growth and potential yield. Knott (1994) demonstrated that a system where the seed was broadcast on to the soil surface and then ploughed in early in the spring was detrimental to establishment and yield, but drilling with a conventional seed drill provided a better return due to lower seedbed losses and higher yield. However, the yield declines with late spring drilling due to susceptibility to dry soil conditions later in the year. Several reviews of optimum sowing dates for spring-sown beans in different countries were summarized by Hebblethwaite *et al.* (1983), with most countries opting to plant as early in the spring as soil conditions allow.

Much work has been carried out on evaluating an optimum population for spring beans. Ingram and Hebblethwaite (1976) concluded from experiments that populations of up to 86 plants/m^2 gave the highest yield but that this excluded the cost of the seed. Work by Cleal (1991) reported that the economic optimum for spring beans was 42–60 plants/m^2. This has subsequently been

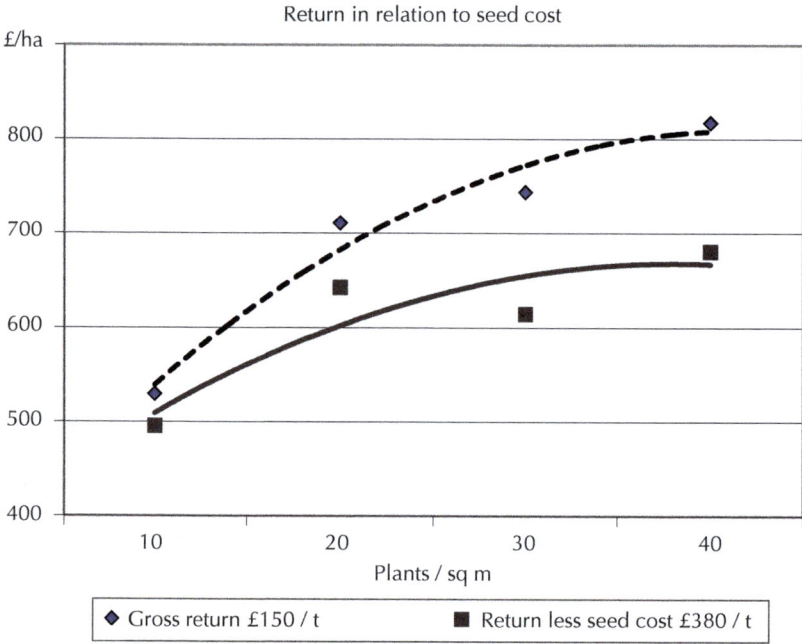

Fig. 4.1. Economic response of winter faba beans to increasing plant density (Armitage, 2009).

confirmed with the newer compact spring varieties (Belcher, 2014). Where comparisons have been made with spring beans planted at decreasing row widths, yields were increased with narrower rows by around 0.35% per centimetre as row width decreased from 53 cm to 18 cm. Other work also showed that narrow rows (18–24 cm), particularly at high seed rates, gave the highest yields with increases of 0.42% per centimetre as row width decreased from 60–75 cm to 12–18 cm. Work in 1986–1988 comparing row widths of 12 cm and 48 cm showed an average yield benefit of 0.15 t/ha, but also showed that effects were greatest when growth was poor and on shorter-strawed varieties (Knott *et al.*, 1994).

Broad beans

For the freezing crop, most beans are planted in row widths compatible with the drill used. The seed is large and the number of commercially available drills suitable for handling a large seed is limited. Drills with partial vacuum seed metering are usually used for broad beans. The drill is equipped with individual units mounted along a toolbar. On each unit, the seed is delivered into a furrow made by a set of discs and then pressed with a following press wheel. The usual row width is 45 cm and the distance between seeds is about 10 cm, with the object of producing a plant population of 18–20 plants/m^2 (Gane *et al.*, 1975). Growing on wide rows is also suitable for the fresh market crop, as access for picking is made between the rows.

Phaseolus beans for fresh market and processing

Green beans are highly responsive to changes in plant density. Mechanical harvesters are able to pick beans from across several rows and although beans can be grown on a bed system with each 1 m wide bed containing four rows, they can be planted without a bed system. In general the row width is 50–90 cm and seed spacing around 5–10 cm apart. Seed is usually drilled at a depth of 5–10 cm, depending on soil type and soil moisture. The optimum population is 60 plants/m^2

Phaseolus beans as dried beans

For large-scale commercial production of dried beans in Canada and other countries where large-scale production is carried out, most are drilled on narrow row widths (50 cm) or wide rows (75–90 cm). The choice of row width depends on the cropping system: narrow rows are favoured with cereal or oilseeds whilst wide rows are favoured by those growing other row crops such as potatoes or maize. The plant population is 17–25 plants/m^2. Inter-row weeding can be carried out while the plants are small. After drilling, the soil is usually rolled to compress soil and stones in order to reduce the soil tare at harvest (Goodwin, 2003).

In small-scale production systems in tropical and subtropical countries, beans may be intercropped with maize. This may follow a number of different systems and combinations, depending on the availability of land and the economics. Relay intercropping of bush or semi-climbing beans planted as soon as the maize is mature is especially common. A second combination is where beans are planted between rows of maize at the same time, though in some areas there is a difference of several weeks between the two harvest periods. Another system is planting beans with other crops such as bananas, cassava, sweet potatoes, coffee and young sugarcane. Each system has been thoroughly described by Woolley *et al.* (1991). Intercropping generally produces low yields, but the advantage to small farmers is the reduction of risk if one crop fails. The different systems demand a range of sowing densities and these have been reviewed by Woolley and Davis (1991).

TIME OF SOWING

Vining and fresh market peas

A succession of vining pea crops must each be harvested at a critical stage of maturity, as determined over a period of around 6 weeks. Two factors in crop production allow this by the use of early- and late-maturing varieties and controlled sowings extending from early spring until early summer. The start of the harvesting season is usually determined by the processor, as there will be requirements for a target production figure for the season and an optimum daily throughput at the factory. Such planning is also required for the large-scale fresh pea market where the provision of adequate labour at harvesting is essential.

This information is used by the individual producer, or producer cooperative, to plan the output for each production area. The object of a sowing programme is for each crop to reach the desired maturity in succession, providing a smooth progression in harvesting and processing, with a product of consistent quality. The intervals between sowings are based on the fact that the major factor affecting growth rate is temperature.

Accumulated heat units (AHUs) may be defined as the difference between the base temperature for crop growth and the mean of the daily maximum and minimum air temperatures. The base temperature is that below which no growth occurs and for peas this is 4.4°C. In the UK, and areas of similar growing conditions, it is usual to allow 11–12 AHUs between each sowing period to provide a difference of 1 day's harvesting between sowings. Thus in cool spring conditions sowings may take place 2–3 weeks apart, whilst later in the season the interval is much shorter. It is usual to use a number of varieties with differing predicted maturity differences, hence the use of early varieties, main crop varieties and late varieties in a programme. These need to be integrated into the AHU programme to allow an overlap of harvesting equivalent to the difference of the maturity date of the varieties (Biddle *et al.*, 1988).

A different approach is used in some areas, where successive sowings are started as soon as the seed in the previous sowing is seen to begin germination, i.e. when the radicle has emerged from the cotyledon and before the plumule is visible. Again the use of varieties with different maturity dates must be integrated in this method.

Combining peas

There is a correlation between the date of sowing of combining peas and yield, though soil conditions in the early spring have a more important influence on the sowing date than the earliness of planting. Early sowing invariably makes a significant contribution towards high yield and this fact was established in the 1950s where yield was found to decrease by 0.125 t/ha for each week's delay after the beginning of March or the first opportunity to plant in the spring (Biddle *et al.*, 1988); however, the precise date of sowing must depend on soil and weather conditions.

Autumn sowing is practised in some countries where it is necessary to establish a crop before the summer temperatures become high and where drought conditions are likely. In these cases, peas should be planted as late as possible in the autumn, when the germination and early seedling development are slower, to avoid wind damage to the newly emerged seedlings. Once established, winter-hardy pea varieties are able to withstand frosts for long periods of cold weather.

Winter faba beans

Most winter beans currently grown in Europe have been bred in the UK and are usually winter hardy, but even these may not survive if conditions are especially severe. However, damage to the main shoot by frost generally stimulates stem branching. Winter beans do not require a vernalization period. The rate of growth depends greatly on temperature, solar radiation and moisture and while very early-sown crops (September) and early-emerged beans may grow to a height of 20 cm by November, neither they nor late-emerged plants make significant amounts of growth until the onset of warmer weather in the spring.

Sowing earlier than mid-October results in advanced vigorous growth and recent research has shown that the beans are more susceptible to chocolate spot (*Botrytis fabae*), *Ascochyta fabae* and the effects of cold wet weather and frost. Although early sowing advances the date of flowering, this does not affect harvest date. Plant height and lodging increase with early sowing. Winter beans sown late into cold wet soil from mid-November to early December suffer from higher seedbed losses. Branching is reduced by late sowing. Sowing after December is usually too late, though recent work has shown that winter beans can be successfully grown if planting is delayed until the spring, which is the normal time for spring varieties (Belcher, 2014).

In several series of UK experiments where beans were sown at different times, the optimum sowing date was obtained where beans were well established plants at the first or second node stage to overwinter. This was normally found in the UK by sowing from mid-October to early November (Knott *et al.*, 1994).

Spring faba beans

Early spring sowings produce the highest yield provided that soil conditions are good, allowing rapid germination and seedling development. To achieve a good

seedbed for spring beans it is normal to plough soil in the autumn to allow winter frosts to weather the soil and to allow a tilth to be produced in the spring
using the minimum of cultivations.

In Europe there is an increasing amount of spring bean production using
non-inversion techniques, either by direct drilling into desiccated stubble or by
minimum cultivations to prepare a level surface before drilling into a disturbed
soil. In any of these systems it is important to produce a weed-free surface and
effective straw dispersal.

Most drilling is carried out as early in the season as soil conditions allow. In
a series of experiments carried out in the UK, yield began to decrease for beans
sown after March. Current work in the UK is examining the relationship
between time of drilling and seed rates but results are not yet conclusive
(Belcher, 2014).

Broad beans

Most broad beans are sown in the spring except for those that are grown for
fresh market, where there is a requirement for beans in early summer and sowings of winter-hardy varieties are planted in the autumn. For the majority of
beans grown for freezing, the harvest period lasts for around 14 days between
the time of harvesting and processing vining peas and before dwarf green
beans. The harvest period is dictated somewhat by factory capacity and, as
with vining peas, harvesting at critical stages is necessary to maximize the
quality of the produce.

It is important that the sowing date is correctly calculated, as this is the
only means of ensuring that the crop is ready for harvest at the correct time.
For freezing, there are relatively few varieties of broad bean in use and of these
there are only small differences in the relative maturity, making the planning
of a sequential harvesting schedule more restricted. The influence of the date
of sowing on the time of maturity of broad beans was investigated in the
1970s, when it was established that consistently successful results were
obtained by relating sowing time to anticipated harvesting time, using the
development curve reproduced in Fig. 4.2.

This relationship is accurate for most UK areas, though in recent years
there has been a move northwards for production into Scotland and therefore
there are slight differences in the rate of maturity of crops where broad beans
are sown early and soil and air temperatures may be lower.

Phaseolus beans for fresh market and processing

Green beans are usually harvested over a 6-week season, lasting from mid-
summer to early autumn, and during this stage successive crops are required,

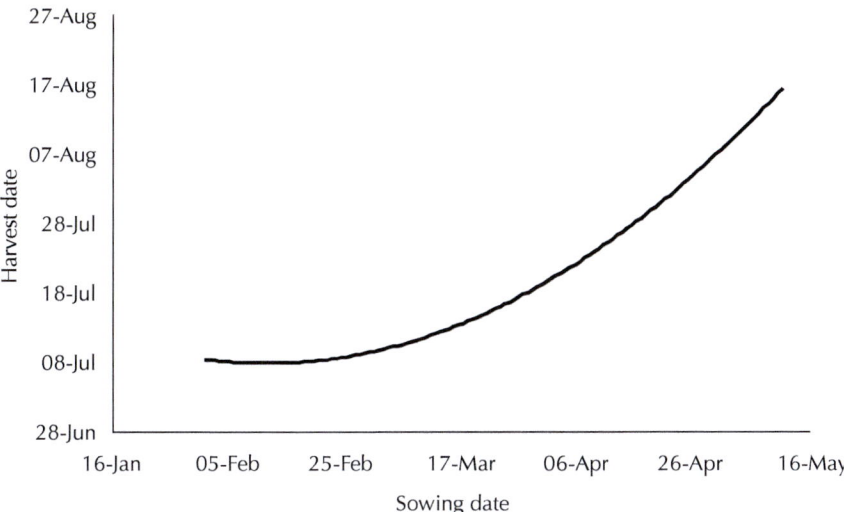

Fig.4.2. Relationship between sowing and harvest date for broad beans for freezing (from Gane *et al.*, 1975).

each at a critical stage of maturity, to give a product of consistently good quality.

The earliness of harvest depends on the market requirements and the climatic conditions of the season and the growing area, notably the time at which the soil temperature reaches 10°C and the likelihood of late frosts. In most growing areas the crop's cold sensitivity makes drilling before then unreliable. Once appropriate soil temperatures have been reached, then sowing can begin.

Because green beans are grown in warmer times of the year, the temperatures at sowing time are often similar to those at harvest; consequently, with sowing and harvesting periods of approximately equal length, sowing programmes can be as simple as sowing 20 days' harvesting capacity every second or third day. Refinements can be made to this concept, such as slightly longer intervals for the earlier drillings and proportionately shorter ones for the later drillings, and the frequency of drilling may be increased during exceptionally warm periods. A system of accumulated heat units can be used for green beans, as with vining peas (see earlier), but the base temperature at which no obvious growth occurs is 10°C. If, for example, the maximum and minimum temperatures of a given day are 18°C and 10°C, respectively, the mean is 14°C and after subtracting the base temperature of 10°C, 4 AHUs are recorded. In order to plan sowings by this method it is necessary to estimate the number of AHUs per day at harvest and then allow that number to accumulate between drillings of 1 day's harvesting capacity, or multiples thereof (Gane *et al.*, 1975). Any sowing programme may contain varieties with different dates of maturity and these differences have to be integrated into the sowing programme.

Phaseolus beans as dried beans

Dried beans have a relatively short growing season; they are shallow rooted and are sown at one of the driest periods of the year. They are particularly sensitive to soil moisture stress and yields are likely to fluctuate when grown without access to irrigation, depending on the weather conditions during growth. As there is little or no discernible growth at temperatures below 10°C, sowing begins as soon as this has been reached and, because beans are very sensitive to frosts, sowing is done after the frost risk has passed. In tropical or semi-tropical production areas, where there is no winter weathering of the soil, cultivations are made shortly before planting and before the onset of the rainy season.

NUTRIENT REQUIREMENTS

Peas

Peas require only a minimal level of fertility. Vining peas have a relatively short growing season but they have a well developed taproot and a limited amount of lateral rooting. Combining peas grow over a longer season and as such can become deficient in some nutrients before maturation. They are less responsive to nutrients in the form of fertilizer than most crops. The major nutrients are potassium (K) and phosphorus (P), whilst nitrogen (N) is supplied through nitrogen fixation in the roots.

Potash (K_2O) is the most important of the major elements; a deficiency has a detrimental effect on the yield as it contributes to the enzyme process involved in nitrogen fixation. The optimum level of soil potassium is around 200 mg/l, above which there is no further yield increase. Severe deficiency causes yellowing and scorching around the leaf margins and, because potassium is associated with nitrogen fixation, nitrogen deficiency can also occur, causing chlorosis and stunting in severe cases. The critical leaf concentration for potassium is below 1.1–1.2% dry matter.

Phosphate (P_2O_5) was thought to have little influence on the growth and yield of vining peas but studies continue to demonstrate the importance of P in vining pea nutrition. Phosphates are an essential nutrient for effective nodulation of pea roots and it is thought that a deficiency may reduce the root nodule number and hence affect the efficiency of nitrogen fixation. A more recent study showed that the response of vining peas to different values of soil P is related to both yield and quality (Morris, 2013). A deficiency of phosphate can reduce seedling vigour, reduce yield and result in precocious maturity at harvest, with the optimum Olsen value of P being around 26–45 mg/kg.

As a general rule, peas are able to make normal growth and give maximum yield through nitrogen fixation in the root nodules. The presence of nitrate (NO_3) in the soil delays nodule establishment and thus impairs nitrogen-fixing activity of the plant. Once nodules are established, when mineral nitrogen is available, it seems that root absorption can complement or replace symbiotic fixation and where N is present at high levels in the soil then natural nodulation ceases. Vining peas are capable of fixing around 200 kg N during the growing season and therefore there is sufficient N to sustain growth while leaving a residue of N in the haulm and root system which then becomes available for a following crop. The nitrogen-fixing bacterium *Rhizobium leguminosarum*, which is responsible for nodulation in peas, appears to be readily available in the soils of the UK, Europe and the Americas. In the absence of a naturally occurring population of *Rhizobium*, inoculation of the seed or soil is practised.

As well as the major advantage of nitrogen fixation from the atmosphere, the impact on the environment is greatly lessened through the reduction in nitrous oxide (N_2O) emissions by legumes in general. Measurements of N_2O taken in field experiments over several years in the UK have shown that N_2O emissions are extremely low throughout the growing period and in the case of vining peas after the vined peas have been removed from the field (R. Sylvester-Bradley, 2014, unpublished). Similar results have been reported by researchers in Europe (Jeuffroy *et al.*, 2013) and Canada.

Peas are also susceptible to deficiencies of some trace elements, particularly manganese (Mn) and sulphur (S). Manganese deficiency occurs mainly on alkaline or organic soils and affects the production of chlorophyll, consequently resulting in interveinal chlorosis. In more mature crops, it is also responsible for a condition known as 'marsh spot' in which the cells in the adaxial surfaces of the cotyledons become necrotic, thereby reducing the quality of the produce and which, in the case of mature crops, can reduce the germination capacity of the resulting seed.

Manganese deficiency is most likely to occur on high-organic soils and is also encouraged by high alkalinity, particularly where soils have been over-limed; and it is more likely to occur when the plants are under stress, such as in exceptionally dry seasons, very wet seasons and where there is soil compaction. Correction of deficiency and prevention of marsh spot are through repeat foliar applications of manganese sulphate (Knott, 1996).

Sulphur deficiency can occur where peas are grown on very light free-draining soils. Deficient crops are stunted and pale in colour, which in turn adversely affects yield. Although the response to sulphur is not as great as in some of the brassica crops such as oilseed rape (canola), pre-drilling soil applications of elemental sulphur equivalent to 25–35 kg SO_3/ha can be beneficial where sulphur levels are known to be low (PGRO, 2015).

Peas are moderately tolerant of acidity, requiring a pH in the region of 5.9–6.5. Liming is necessary for soils of pH lower than 5.8 or the crop will be very stunted and pale and root nodulation inhibited or absent. They will grow normally in soils of higher pH provided that the effects of manganese deficiency are rectified.

Deficiency of magnesium (Mg) is less common than of manganese, especially in the UK. It occurs late in the growing season and is seldom severe enough to warrant remedial treatment. Magnesium is an essential element for chlorophyll production and a deficiency causes interveinal chlorosis but, unlike in the case of manganese deficiency, the margins of the leaf remain green. It does not affect the quality of the produce, though there is some evidence that yield may be lost in exceptional circumstances. It is most prevalent in light soils in wet seasons and can be induced on many soils by excessive applications of potash.

In general, fertilizer is applied to peas depending on the levels of the major nutrients in the soil. Fertilizer recommendations vary slightly between countries but Table 4.2 shows the fertilizer requirements for peas as recommended in the UK.

Faba beans

Soil pH is an important factor affecting the yield of *V. faba* and optimum pH before sowing is about 6.9–7.3. However, in many drier parts of the world pH values of 8.5 are common and this does not generally appear to affect the crop adversely, provided that sufficient micronutrients are available. Crops grown on acid soils are less vigorous, paler in colour and lower yielding than crops grown at the optimum pH ranges. The loss in vigour results in smaller seeds and a lower weight of seeds per plant.

As with peas, *V. faba* grown for either fresh broad beans or dried faba beans has a limited requirement for fertilizer. Spring beans have a shorter growing season than winter beans but growth of winter beans is slow until the early spring. Beans are less responsive to fertilizer application than many other crops but deficiencies can occur in some soils due to low fertility, soil compaction or acidity.

As a legume, *V. faba* is supplied with its nitrogen requirements by nitrogen-fixing rhizobia. Rhizobia are widespread in agricultural soils, including most

Table 4.2. The fertilizer requirements of peas (kg/ha) (Defra, 2010).

Soil index[a]	N	P_2O_5 (kg/ha)	K_2O (kg/ha)	MgO (kg/ha)
0	0	100	100	100
1	0	70	70	50
2	0	40	40 (2–)	0
			20 (2+)	
>2	0	0	0	0

[a]According to soil analysis on the ADAS classification: 0 = very low; 1 = low; 2 = medium; >2 = high

UK soils, though they may not be present in areas where there has been no history of *V. faba* growing and then rhizobia are often applied as an inoculant, either to the soil pre-planting or as a seed application.

Applications of soil nitrogen have little effect on yield and beans have a poor utilization efficiency, though some increases in vegetative growth may be obtained. Excessive soil nitrogen can inhibit nodulation.

Phosphate levels in most agricultural soils are relatively high and very little response to phosphate is gained by large applications. Beans are more responsive to potassium and low potash can result in lower numbers of root nodules. There is some evidence that low potash can predispose *V. faba* to infection of chocolate spot (*Botrytis fabae*). The fertilizer requirement of *Vicia* beans is shown in Table 4.3.

Few minor nutrients are of significance but of those that do have an effect on growth or yield, correction is usually by means of foliar treatment. Deficiency or unavailability of manganese adversely affects chlorophyll production and may cause chlorosis. It is also responsible for marsh spot disorder, which reduces the quality of the seed or beans for human consumption. It is most likely to occur on highly organic, alkaline or heavily limed soils and when plants are under stress through drought, excess water or compacted soils. Faba beans are much less sensitive to manganese deficiency than peas and symptoms are rarely seen.

Magnesium deficiency is also rare, except on acid soils. The symptoms are interveinal chlorosis but with the leaf margins remaining green. Often symptoms of nutrient deficiency can be confused with aphid-transmitted viruses (see Chapter 5).

Boron deficiency occurs occasionally on some soils and is best corrected with boronated fertilizer. Symptoms include interveinal chlorosis, starting at leaf margins, and young leaves may be cupped or deformed. The stems may develop shortened internodes and all symptoms can occur together.

On alkaline soils in Australia, crops have been shown to respond to zinc; and molybdenum can be applied as a soluble fertilizer to acid soils.

Table 4.3. The fertilizer requirements of faba beans (Defra, 2010).

Soil index[a]	N	P_2O_5 (kg/ha)	K_2O (kg/ha)
0	0	100	100
1	0	75	70
2	0	40	40
>2	0	0	20

[a]According to soil analysis on the ADAS classification: 0 = very low; 1 = low; 2 = medium; >2 = high

Phaseolus beans

Phaseolus beans grow best on neutral soils with a pH range of between 6.0 and 8.0. Most South American and African soils are acidic, high in aluminium and manganese and low in phosphate, and pH correction by liming is normally practised as beans are susceptible to aluminium toxicity.

In general it is thought that *Phaseolus* beans are a low-potential nitrogen-fixing crop and in many experiments beans have responded well to high applications of nitrogen. Although rhizobia are present in soils, particularly in South America, often the level of nitrogen fixation is not enough to maximize the potential of the crop throughout its relatively short life. It is thought that there is a wide range of strain-specific rhizobia present in some soils and specific interactions between strains and cultivars may be significant. In North American dried bean production, the use of a rhizobium inoculant is not practised and nitrogen is therefore applied as a fertilizer prior to planting. In general, US growers apply between 150 and 250 kg nitrogen/ha before planting, either in the spring or the previous autumn. There has been some development with soil-applied granular inoculants with which, when applied with the seed at drilling, the amount of nitrogen fertilizer can be reduced (Biddle, 2009).

Phosphorus deficiency is most commonly found in all acid soils of the world. The plant remains dwarfed and the stems are thin with short internodes. The upper leaves remain small and dark green while the lower leaves become yellow with necrosis at the edge.

Potassium deficiency causes yellowing and necrosis of the leaf tips and margins but symptoms are not often seen in the field as other factors, such as lack of moisture or excessive water, can often mask the symptoms. It is normal practice to apply compound fertilizers of N, P and K together. The rates commonly used are shown in Table 4.4.

Beans often are affected by a deficiency or toxicity of micronutrients. Sulphur deficiency appears as the whole leaf turns chlorotic in the lower part and later affects the upper leaves. This is particularly a problem where compound fertilizers are used that have little or no additional sulphur. Zinc deficiency

Table 4.4. The fertilizer requirements of beans (Defra, 2010).

Soil index[a]	N (kg/ha)	P_2O_5 (kg/ha)	K_2O (kg/ha)
0	180	200	200
1	150	150	150
2	120	100	50–100
3	80	50	50

[a]According to soil analysis on the ADAS classification: 0 = very low; 1 = low; 2 = medium; 3 = high

causes interveinal chlorosis of young leaves and necrotic spotting occurs on the leaf surface. Zinc is particularly deficient in alkaline soils in some American countries and Australia. Boron deficiency normally occurs shortly after germination when the bean is still small. In severe instances the growing point dies and a proliferation of secondary buds occurs. Manganese toxicity occurs in soils of volcanic origin with pH of less than 5.5 and can also be induced by excess ammonium sulphate application.

WATER REQUIREMENTS

Vining and fresh market peas

It is well known that water deficit is a major limiting factor in peas (Pumphrey *et al.*, 1979).The most critical stages of pea plant growth occur when water deficit coincides with the vegetative growth and up to flower initiation. Growth is reduced and the size of flowers and pods is smaller. If water deficit occurs between the beginning of flowering and the final stage of pod development, the number of seeds set in the pods is reduced. Work has shown the positive effects of moderate water deficit on yield when occurring shortly before and after flowering (Salter, 1962; Stocker, 1973). Recent studies in the UK compared growth and yield of peas under drought stress conditions with those maintained on soils at field capacity and found that growth and seed yield and seed weight were dramatically reduced in droughted peas (E. Uber and A.J. Biddle, 2009, unpublished). The response to irrigation is therefore very significant in dry conditions (Fig. 4.3).

Timing of irrigation is crucial and should be related to specific growth stages and soil moisture deficit. Soils are usually near to field capacity when most peas are sown and irrigation rarely increases yield if applied before the start of flowering, though it does increase haulm. Irrigation during vegetative growth can depress pea yields and is therefore not used unless the seedbed is very dry and adequate germination would not otherwise occur or where the crop is severely wilted.

Peas are most responsive to irrigation when first flowers are opening. It is thought that the response is greatest at this stage because the root system has ceased to grow, making the plant more vulnerable to water shortage. Yield increases from irrigation at this time are often very substantial and may be up to 50%, due to a greater number of pods contributing to yield and more peas per pod.

Irrigation at late flowering or petal fall does not result in yield increase, there being no effect on either the weight or number of pods per plant or on the weight of peas per pod. A slight renewal of root growth occurs during this period but no increase in haulm weight. Irrigation at petal fall may increase the occurrence of pod rot caused by *Botrytis cinerea*.

Fig. 4.3.   Pivot irrigator in vining peas.

Irrigation applied at pod swell shows significant yield increase. The number of peas per pod and the mean weight of peas are both increased. An indication of the yield response of vining peas is shown in Table 4.5.

Irrigation during vegetative growth and early flowering has been found to have little effect on the rate of maturation, but applications during pod swelling can delay the harvest date of peas for freezing or canning by about 2 days (HDC, 2012).

Table 4.5.   Response of vining peas to irrigation (Biddle *et al.*, 1988; data summarized from Institute of Horticultural Research, Wellesbourne, UK).

Growth stage at irrigation (25 mm water/ha)	% Increase in haulm weight	% Increase in shelled pea weight
Vegetative growth	60	−5
Onset of flowering	30	30
Pod set	0	0
Pod swell	0	20
Pod set and pod swell	30	40

Combining peas

Work in the UK and elsewhere suggests that the responses are very similar for combining peas and that there is little evidence to suggest that irrigation regimes would need to be different from that shown for vining peas in Table 4.5. A review of the physiological effects of water stress and responses to water has been provided by Lecoeur and Guilioni (2010).

Faba beans

As with peas, beans respond well to irrigation in the presence of a soil moisture deficit. One of the most limiting factors in yield production in spring-sown beans is moisture stress at flowering onwards and this has limited production of spring beans in several areas of the world where the growing season is dominated by drought or lack of irrigation. The main effects of moisture stress are a reduced development of vegetative growth, poor seed set and a reduction in the number of reproductive nodes due to either flower abortion or seed shrivelling. Where drought conditions occur late in the growing season when the pods are filling, the pod walls can degrade and liquefy, producing immature blackened ovules. Field beans are shallow rooting compared with winter cereals, which extract moisture from more than 1 m depth. The main roots of beans can extract water from 40 cm and the sparser rooting system can extract easily available water (spring beans at 70 cm and winter beans at a depth of 90 cm). Several workers have shown yield and growth responses to irrigation of beans but generally the larger responses have been obtained with spring beans rather than winter beans, probably because the root systems of autumn-planted beans have developed strongly and extensively through the soil profile, mitigating the effects of dry soil in the upper rooting zone (Husain *et al.*, 1988).

The response of spring faba beans to irrigation applied before, during and after flowering and in combinations was studied over 3 years on sandy soils in the UK by Knott (1999). Irrigation increased spring field bean yield in all three years, even in a year when a post-flowering period of drought caused a yield depression of 2 t/ha on non-irrigated plots. The yield response for fully irrigated compared with non-irrigated field beans per 25 mm water was 0·34, 0·28 and 0·36 t/ha in each year, respectively. Irrigation applied during or after flowering gave statistically significant yield increases and there were indications of greater efficiency of water use from post-flowering applications.

Phaseolus beans

Dried beans in particular are grown in a very wide area of the world, many of which have periods of low rainfall and high temperatures that can exacerbate the effects of drought on bean production. In many developing countries, irrigation is not always available at an economically viable cost although small producers have developed systems that utilize ridges and furrows to allow irrigation channels to provide water when required. On large-scale production systems in areas such as Central America, where beans are often planted towards the end of the rainy season, initial soil moisture is usually adequate and the stress period depends on when the rains cease. Beans in more northern countries of South America are planted at the onset of the short unreliable rainy seasons, but production problems associated with this variability of rain may exacerbate drought effects such as salinity, high temperatures, pathogenic root-infecting fungi and insects.

In North America, cropping on a large scale is made by planting as early as possible in the spring, relying on the presence of sufficient moisture in the soil to maintain early growth and then reliance on rainfall to supply enough moisture through to crop maturity. However, dry beans require an average of 380 mm of moisture through the growing season. This can be spring soil moisture, in-season rainfall and/or irrigation. The critical water requirement period for dry beans is from late bud through to pod formation. Under average climatic conditions, bean daily water use peaks at 6–7 mm but beans can utilize over 7.5 mm per day when temperatures are over 30°C. Applications of irrigation after drilling and before emergence cool the soil and may delay emergence. In Canada there is a reliance on the use of varieties that have been bred for their short season and are less likely to suffer from drought stress. Research has shown that yields increase under irrigation when soil moisture is adequate through the vegetative period as well as through flowering and pod formation. During the vegetative period, the active root zone is 30 cm deep. While plant requirements for moisture are not high during this time, early stress can reduce the number of branches and sites for podding. During the reproductive period, the active root zone is 80 cm deep. Adequate moisture during this period is critical for maintaining yield. However, irrigating frequently and maintaining moist surface soil conditions during pod formation increases the risk of infection by *Sclerotinia sclerotiorum*, causing white mould. Building up and maintaining moisture to a depth of 80 cm before flowering allows for a reduction in irrigation frequency during pod formation and pod filling. Irrigation can then continue as required until the crop has nearly reached full maturity (Shaw, 2009).

Beans grown for processing have similar water requirements. As the beans may be planted in succession, the requirement for irrigation is more important later in the season, necessitating irrigation equipment that can be easily moved across or between fields.

SUMMARY

Each legume crop requires its own level of management to ensure rapid successful germination of seed and the differences in seed characteristics should be considered before calculating a sowing rate and drilling at the appropriate time for each crop. Harvesting as vegetables, particularly where the produce is processed, requires a succession of planting to provide a constant input to the factory over the processing season. Each crop species has a different set of criteria for planning a planting schedule but the principles are very similar between species. Although peas and beans have little requirement for the addition of nitrogen fertilizer, the use of inoculants for crops grown in soils with an absence of rhizobia is a further consideration. Responses to other major nutrients may be relatively small but maximum production can only be achieved by efficient use of nutrients where these are in short supply in the soil. Peas and beans are very drought sensitive and water requirements and applications either by natural means or by irrigation can be critical.

REFERENCES

Armitage, P.A. (2009) Seed rates – getting it right. *The Pulse Magazine*, Spring 2009. Processors and Growers Research Organisation, Peterborough, UK, p. 8.

Belcher, S.J. (2014) Pulse agronomy. *The Pulse Magazine*, Spring 2014. Processors and Growers Research Organisation, Peterborough UK, p.4.

Biddle, A.J. (1983) The foot rot complex and its effect on vining pea yield. *Plant Protection for Human Welfare: Proceedings of 10th International Congress of Plant Protection, Brighton, 20–25 November 1983*, 1, 117. British Crop Protection Council, Farnham, UK.

Biddle, A.J. (2009) *Dwarf Green Beans: Evaluation of Rhizobium Inoculant for Nitrogen Fixation*. Report FV354. Agriculture and Horticulture Development Board (AHDB), Kenilworth, UK.

Biddle, A.J., Knott, C.M. and Gent, G.P. (1988) *The PGRO Pea Growing Handbook*. Processors and Growers Research Organisation, Peterborough, UK.

Bladon, F.L. and Biddle, A.J. (1991) A three year study of laboratory germination, electrical conductivity and field emergence in combining peas. In: *Aspects of Applied Biology 27, Production and Protection of Legumes*, pp. 305-308. Association of Applied Biologists, Warwick, UK.

Cleal, R.A.E. (1991) The effect of plant population on the yield of semi-determinate spring field beans. *Aspects of Applied Biology 27, Production and Protection of Legumes*, pp. 89–94.

Defra (2010) *Fertiliser Manual (RB209)*, 8th edition. Department for Environment and Rural Affairs, London.

Gane, A.J., King, J.M. and Gent, G.P. (1975) *PGRO Pea and Bean Growing Handbook – Beans.* Processors and Growers Research Organisation, Peterborough, UK, pp. 35–37.

Goodwin, M. (2003) Crop profile for dry beans. *Pulse Canada*, February 2003.

HDC (2012) *Irrigating Vining Peas.* Fact Sheet 02/12. Agriculture and Horticulture Development Board (AHDB), Kenilworth, UK.

Hebblethwaite, P.D., Hawtin, G.C. and Latman, P.J.W. (1983) The husbandry of establishment and maintenance. In: Hebblethwaite, P.D. (ed.) *Faba Bean (Vicia faba L.). A Basis for Improvement.* Cambridge University Press, Cambridge, UK, pp. 271–312.

Husain, M.M., Hill, G.D. and Gallagher, J.N. (1988) The response of field beans (*Vicia faba* L.) to irrigation and sowing date. *Journal of Agricultural Science* 111(2), 233–254.

Ingram, J. and Hebblethwaite, P.D. (1976) Optimum economic seed rates in spring and autumn sown field beans. *Agricultural Progress* 51, 27–32.

ISTA (2006) *International Rules for Seed Testing.* Edition 2006. International Seed Testing Association, Bassersdorf, Switzerland.

Jeuffroy, M.H., Baranger, E., Carrouée, B., de Chezelles, E., Gosme, M. *et al.* (2013) Nitrous oxide emissions from crop rotations including wheat, oilseed rape and dry peas. *Biogeosciences* 10, 1787–1797.

King, J.M. (1967) Vining peas – plant populations and profitability. *Agriculture* 74, 167–170.

Knott, C.M. (1994) Spring bean establishment. *Pea and Bean Progress*, Spring 1994. Processors and Growers Research Organisation, Peterborough, UK.

Knott, C.M. (1996) Control of manganese deficiency in field peas grown for seed or human consumption. *Journal of Agricultural Science* 127(2), 207–213.

Knott, C.M. (1999) Irrigation of spring field beans (*Vicia faba*): response to timing at different crop growth stages. *Journal of Agricultural Science* 132(4), 407–415.

Knott, C.M., Biddle, A.J. and McKeown, B.M. (1994) *The PGRO Bean Growing Handbook.* Processors and Growers Research Organisation, Peterborough, UK.

Lahoud, R., Mustafa, M. and Shehadeh, M. (1979) Food legume production and improvement in Lebanon. In: Hawtin, G.C. and Chancellor, G.J. (eds) *Food Legume Improvement and Development.* ICARDA-IDRC, IDRC-12c, Ottawa, Canada, pp. 69–70.

Lecoeur, J. and Guilioni, L. (2010) Influence of water deficit on pea canopy functioning. In: Munier-Jolain, N., Biarnes, V., Chaillet, I., Lecoeur, J. and Jeuffroy, M.H. (eds) *Physiology of the Pea Crop.* CRC Press, Boca Raton, Florida, pp. 135–147.

Maguire, K. (2013) Maintaining healthy pulse crops. *The Pulse Magazine*, Spring 2013. Processors and Growers Research Organisation, Peterborough, UK.

Matthews, S. (1971) A study of seed lots of peas (*Pisum sativum* L.) differing in predisposition to pre-emergence mortality in soil. *Annals of Applied Biology* 68, 177–183.

Morris, N. (2013) *Identification of critical P in vining pea crops.* HDC Project FV380 Annual Report 2013. Agriculture and Horticulture Development Board (AHDB), Kenilworth, UK.

PGRO (2014) *Electrical conductivity test for vining pea seed.* Technical Update 35, January 2014. Processors and Growers Research Organisation, Peterborough, UK.

PGRO (2015) *PGRO Pulse Agronomy Guide*. Processors and Growers Research Organisation, Peterborough, UK.

Pumphrey, F.V., Ramig, R.E. and Allmaras, R.R. (1979) Field response of peas (*Pisum sativum*) to precipitation and excess heat. *Journal of the American Society of Horticultural Science* 104(4), 548–550.

Salter, P.J. (1962) Some responses of peas to irrigation at different growth stages. *Journal of Horticultural Science* 37, 141–149.

Shaw, L. (2009) *Management of Irrigated Dry Beans*. Irrigation Factsheet. Saskatchewan Ministry of Agriculture, Saskatoon, Canada.

Smith, K.B. (2007) *Vining peas: Evaluation of precision sowing*. Report FV297. Horticultural Development Council, Kenilworth, UK.

Stocker, R. (1973) Response of viner peas to water during different phases of growth. *New Zealand Journal of Experimental Agriculture* 1, 73–76.

Villalobos, M.J.P. and Jellis, G.J. (1990) Factors influencing establishment in *Vicia faba*. *Journal of Agricultural Science* 115, 57–62.

Woolley, J. and Davis, J.H.C. (1991) The agronomy of intercropping with beans. In: van Schoonhoven, A. and Voysest, O. (eds) *Common Beans – Research for Crop Improvement*. CAB International, Wallingford, UK, pp. 707–735.

Woolley, J., Idefonso, R.L., de Aquino Portes e Castro, T. and Voss, J. (1991) Bean cropping systems for the tropics and subtropics and their determinants. In: van Schoonhoven, A. and Voysest, O. (eds) *Common Beans – Research for Crop Improvement*. CAB International, Wallingford, UK, pp. 670–706.

MANAGEMENT OF WEEDS

Large-seeded legumes such as peas and beans, compared with many other agricultural crops, do not offer very great competition to weeds and consequently infestations can cause yield depression. Weed control has become more efficient in recent years, with the introduction of new pesticides, but this situation is becoming more difficult with the increasing emphasis on reducing pesticide usage and reducing the number of active ingredients believed to be detrimental to the environment (Grundy *et al.*, 2011).

In many countries where the crops are grown on a small scale, access to pesticides is very limited because of availability or economics, therefore reliance on the use of such materials may not be possible or indeed sustainable. There are a number of cultural aids to weed control that can be utilized, increasingly so in large-scale commercial production, which can eliminate the need for chemical weed control or at least reduce its frequency of use. Weeds have a long-term effect on crop rotation, such as the return of seeds of the grass weeds *Avena fatua* and *Alopecurus myosuroides* and the spread of perennial weeds in succeeding crops. Weed flora is dependent on climate, soil type, crop rotation and time of sowing and no weeds are specific to food legumes, with the exception of the parasite *Orobanche* spp. (especially *O. crenata*).

The first part of this chapter discusses the common methods of weed control in peas and beans with the use of herbicides and where cultural control is available.

WEED PROBLEMS

Peas

The competitive ability of the crop is enhanced by an evenly distributed plant population, giving good ground cover, in narrow rows. Although this will not be sufficient to smother weeds completely, it will increase the effectiveness of weed control measures that may need to be applied. Weeds not only reduce yields, they may also seriously affect the handling of the crop and the quality of the harvested peas (Knott and Halila, 1988). The harvesting of vining peas by self-propelled viners can be seriously hampered by weeds such as *Polygonum* and *Stellaria* spp. in Europe, sometimes causing blockages. Dense populations of weeds also slow down throughput within the viner and efficiency of recovery of peas is severely reduced by the bulk of the weeds passing through the drum, extending the harvest time and risking the quality of the vined peas. In combining peas, weed growth slows down the natural drying of the crop and increases the time that it is exposed to the weather and disease, thereby reducing quality and profit. Weeds also slow down the rate of combining and create difficulties. Chemical desiccation may be necessary to kill the weeds and this adds to production costs (Knott, 2002).

In vining peas, contamination of the produce with weed debris creates problems for the processor. Seed heads of thistle (*Cirsium arvense*) and common poppy (*Papaver rhoeas*) and mayweeds (*Matricaria* and *Tripleurospermum* spp.), the berries of black nightshade (*Solanum nigrum*) and white bryony (*Bryonia dioica*) and volunteer potatoes can all be extremely difficult to separate from the produce. Although factories are able to remove contaminants of a different colour from peas, those of the same colour and specific gravity as peas remain. If the level of contamination is high when the peas reach the factory, a percentage may not be removed by the normal cleaning operations. With the limited opportunity available for hand sorting in most factories, the only solution is to reject the produce from the infested area.

Apart from contamination, some weeds may taint the produce. For example, stinking mayweed (*Anthemis cotula*) and berries of black nightshade, bryony and potato are to some extent toxic and although it is unlikely that sufficient quantity would be present in a sample to create serious consumer hazard, the risk is still there. Some weeds are hosts of crop pathogens and their presence increases the risk of diseases developing in the crop; for example, some cruciferous weeds are hosts of white mould (*Sclerotinia sclerotiorum*).

Grass weeds are particularly a cause of yield loss in peas through direct competition. Some annual weeds such as wild oat (*Avena fatua*) or black-grass (*Alopecurus myosuroides*) can develop large seed banks that can remain viable for many years; in addition, repetitive use of certain herbicide active ingredients in the farm rotation have caused populations of some species to develop resistance. Perennial grass weeds such as couch (*Elymus repens*) and others are difficult to control, as they emerge and grow at the same time as the pea crop and there are very few selective graminicides available or that are safe to the crop. Volunteer cereals such as barley can also be a competitor to peas in some situations.

Faba beans

In the early stages, faba beans are not very competitive with weeds. Later, the taller bean varieties compete well with most weeds, though climbing species such as polygonums and cleavers or goosegrass (*Galium aparine*) can still be a problem. In experiments in the UK in spring field beans, where weed infestations ranged from 79 to 157 plants/m^2, yield increases achieved where broadleaved weeds were controlled by pre-emergence herbicides ranged from 13% to 44%, with an average yield response of 21% (Knott *et al.*, 1994). This more than covered the cost of herbicide and application. For world-market prices, 'break-even' yield is about 9%. It is therefore important to control weeds in spring field beans. Yield response was not always adequate to cover the cost of post-emergence herbicides for broadleaved weeds. In terms of yield alone, weeds must be controlled in spring beans but in a series of other experiments, yield increases from weed control were not achieved where weed populations were below 50 plants/m^2. However, since most weed control is practised with pre-emergence products, the weed population is not known at the time of application.

Phaseolus beans

Young bean plants are not competitive and severe yield losses will occur even from low weed pressure. In some situations weeds that build up in previous crops, particularly where beans follow maize, may increase populations of slugs, which attack the following bean crop. Spores of bean rust (*Uromyces phaseoli*) build up on *Oxalis* spp., a common broadleaved weed in bean fields in South America. Weeds may also act as hosts of bean golden mosaic virus and its vectors. Poorly weeded crops suffer more virus infection and develop inadequately, thus permitting further weed growth. In large-scale production, perennial weeds such as creeping thistle or Canada thistle (*Cirsium arvensis*), sow thistle (*Sonchus arvensis*) and couch grass or quack grass (*Elymus repens*) cannot be controlled by chemicals within the crop, so high pressure of these weeds will influence whether or not beans are grown on the field. Other weeds are more difficult to control in the crop, such as wild buck wheat (*Falopia convolvulus*) and volunteer oilseed rape (canola) or flax. Green weeds or weed seeds or berries such as black nightshade (*Solanum nigrum* or *S. americanum*) present at harvest can reduce crop quality through staining of the beans.

In green beans for fresh market or processing, late-germinating weeds such as chickweed (*Stellaria media*), which forms a dense mat, or those that intertwine along the rows such as black bindweed (*Falopia convolvulus*) and knotgrass (*Polygonum aviculare*) seriously interfere with harvest. Tall-growing weeds such as fat hen (*Chenopodium album*) and nettle (*Urtica urens*) probably cause fewer harvesting problems but have a greater effect on yield. Mechanical harvesting of green beans is more likely to result in crop contamination with weed stems and flower heads but volunteer crops, particularly potatoes, pose a risk of contamination with potato berries as well as stalks which are very difficult to remove.

CULTURAL AIDS TO WEED CONTROL

Peas

Many weed problems may be reduced by stubble cleaning, linked with efficient ploughing, when carried out in the autumn. Stubble cleaning is a cheap and efficient way of giving some control of perennial grasses but should take place in dry conditions. Early light cultivation of moist soil after the previous cereal harvest encourages germination of shed cereal or oilseed rape seeds and also encourages germination of annual grasses such as barren brome (*Anisantha sterilis*) but reduction of black-grass and shed wild oats is greatest where stubble is undisturbed and there is predation by birds.

Time of sowing can also influence populations of weeds and later sowings can allow time for a stale seedbed approach. In this technique, cultivated land is allowed to be populated with germinating weeds and then cultivated again just before drilling, disturbing the roots of the weed seedlings and allowing them to desiccate (Davies and Welsh, 2002).

Only light spring cultivations are normally needed to produce a suitable seedbed but cultivations have limited effect where overwintered weeds are well established. In peas, the use of narrow row widths has eliminated the use of inter-row cultivations but the use of tined weeders such as the Einbock weeder, which combs through the top layer of soil, can be effective in removing small weed seedlings without undue damage to the pea plants. Timing is important and work has shown that such weeding can be effective at certain growth stages of the weed and the crop. Early experiments have shown that some control of shallow-rooted annual weeds can be gained by using the weeder at early growth stages, from the second node to the fifth vegetative node. Such treatment is more effective on lighter soil types and should be made when the surface is dry. Passes with the weeder may be made along or across rows without causing significant damage to the crop. Later passes can be made along the rows but may cause damage to pea crops, especially when they have begun to flower. Severe damage will be sustained if passes are made across the rows at later growth stages (PGRO, 2015a).

Faba beans

In bean production, stubble cleaning and stale seedbed techniques are usually employed, though these do not reduce the populations of weeds growing from depth, such as volunteer oilseed rape and charlock (*Sinapis arvensis*). Mechanical cultivation in the crop has the disadvantage of not controlling the weeds within the row and may require more than one operation. Harrows are often used in winter beans and if the growing points are damaged, plants will compensate by producing basal branches. However, beans must be grown in wide rows (40 cm) to facilitate passage between them. The frequency of cultivation required to give adequate weed control is dependent on crop sowing date, number of species of weeds present and growing conditions. In spring beans, early removal by cultivating 4–6 weeks after crop emergence is essential. More frequent cultivations may be necessary in winter beans, as weeds have longer to germinate. Tine weeding with an Einbock type of finger weeder can also be utilized, particularly in spring beans before they grow too tall.

Phaseolus beans

In *Phaseolus* beans, which are often grown on wider rows than peas, mechanical weed control can be undertaken at several stages during the crop's growth. The critical time for weeding is between 3 and 6 weeks after planting, to maintain maximum yield (Burnside *et al.*, 1998). Using a stale seedbed technique first, with irrigation if necessary, will promote early germination of weed seedlings, which can then be lightly cultivated. Inter-row cultivations made after planting should be carried out using a hoe at a depth that will not disturb the developing bean roots.

Weeds are less of a problem in intercropping or mixed cropping. Where beans are grown with maize, the crops fill more than one ecological niche and thus compete effectively with a greater number of weed species. Where beans follow maize, growers take great care in weed control in the maize to reduce the amount of weed-seed shedding before beans are planted (Woolley and Davis, 1991).

BIOLOGICAL CONTROL OF WEEDS

There have been several instances where biopesticides have been used in weed control in certain cropping situations. Such methods may involve the introduction of foreign and native organisms that attack in their natural environment and once established within a crop have had some beneficial effects. However, few such methods have been developed successfully on a large scale and no research has been done on the weed species that affect peas and beans.

Biofumigants

Biofumigants have been used for many years and there continues to be significant research effort in utilizing the techniques in annual crops. The technique is based on the incorporation of fresh, mulched plant material into the soil, which will release several substances able to suppress soil-borne pests or diseases and in some cases there have been claims that weeds are also suppressed. Brassicas are particularly active sulphur accumulators and synthesize significant quantities of sulphur-rich glucosinolates. Damaged leaves release myrosinase enzymes that break down the glucosinolates into several products, including isothiocyanates (which are highly toxic). Work has shown that these products are active against some soil-borne diseases and pests but there are claims that weed suppression can also be obtained.

Growing a brassica cover crop for use as a biofumigant involves the establishment of an autumn-sown crop, usually mustard and in particularly the brown mustard (Caliente) species (*Brassica juncea*), though other members of the Cruciferae may also be used.

Whilst the use of such cover crops has several benefits in arable cropping, such as providing a source of nutrients for newly developing crops, an aid to improving soil structure by adding to the organic matter and some effects on suppressing pests such as nematodes, there is little firm information on the technique becoming a sustainable means of suppressing weeds in peas and beans. It has been suggested that Cruciferae may have an allelopathic effect on weeds and reports from Europe and North America have also suggested that brassicas can be used for integrated weed management (Tollsten and Bergstrom, 1988; Turk *et al.*, 2005). However, there are no reliable data available showing reductions in weed numbers in the growing crop (Haramoto and Gallandt, 2005a,b). Similarly, there are reports of suppression of black-grass (*A. myosuroides*) using a mustard cover crop preceding spring-sown faba beans (Jim Scrimshaw, personal communication), but effects have not been repeatable in beans from year to year, though other work in winter cereals has shown some significant suppression. This may be due in part to a lack of persistence of the glucosinolates in the soil during the germination period of the weed plants or there may be other factors involved.

CHEMICAL WEED CONTROL

In most areas where peas are grown on a large commercial scale, chemical herbicides are still available and effective in a wide range of conditions. Herbicides used for large-seeded legumes can be grouped into three main categories: pre-drilling herbicides; pre-emergence herbicides; and post-emergence herbicides.

Pre-drilling herbicides

These materials are applied before sowing and are often incorporated into the seedbed by cultivation. The cultivations required to mix them into the seedbed efficiently can produce too fine a tilth, which may be detrimental to crop growth. Drilling may have to be delayed on heavier textured soils until soil conditions are suitable for incorporation techniques to be used. Pre-drilling herbicides are usually applied where populations of perennial grasses or known populations of grass weeds such as wild oats or black-grass are present.

Pre-emergence herbicides

These are applied after drilling but before the crop has emerged. Some materials have a contact action on emerged weeds, some kill weeds through uptake from the soil, others work through a combination of both and most of these interfere with photosynthesis. Materials that rely on uptake from the soil are classed as residual herbicides. They have low solubility and are persistent in soil surface layers but, to be effective, most require moisture soon after application to move them into the soil within reach of the weed seeds and roots. Selectivity is achieved by planting peas below the location and availability of the herbicide in the soil and also in some cases by inherent tolerance of the peas to the herbicide. Effectiveness of residual herbicides is influenced by the amount of clay and organic matter in the soil. Clay particles can adsorb the chemical on to the surface, thus locking up a certain proportion and reducing efficiency. Organic matter acts in a similar way. Residual herbicides are degraded by ultraviolet light if they remain on the surface and bacteria complete the process in the soil. Pre-emergence contact herbicides rely solely on killing weeds that emerge before the crop. Since weed control and crop emergence usually coincide, there is little opportunity to apply these in peas.

Post-emergence herbicides

In the case of materials that are applied post emergence, treatment is carried out when both weeds and crop have emerged. Post-emergence herbicides used in peas act on foliage by translocation alone, or by contact and translocation, and in some cases they have residual action as well. Those that have some contact activity rely on the waxy cuticle of the foliage to cause run-off of the spray and those that are translocated are apparently absorbed by the foliage without adversely affecting its development.

Contact activity
Selectivity of contact-acting foliar-applied herbicides depends in part on differential retention of the herbicide by the crop and weeds. Pea foliage is covered

by large amounts of epicuticular wax that repels spray droplets. The leaf wax of fat hen (*Chenopodium album*), for example, is not as water repellent and is damaged by a greater retention of contact herbicide. Other post-emergence herbicides have some contact action on the peas and where the amount of wax on the pea foliage is low or damaged by wind, hail, frost or blowing sand particles or mechanical injury of some sort, then the effectiveness of the wax cuticle is reduced and serious scorch can occur.

Where post-emergence products are to be used, the amount of leaf wax present on the foliage can be checked before spraying by sampling a plant and immersing it in a container with a 1% solution of methyl violet dye. The plant is then removed and excess dye is gently shaken off. The amount of dye retained is an indication of the areas in which the wax cuticle is damaged or inadequate. This level is assessed and a decision on whether or not to apply the herbicide is made when a satisfactory level of tissue that has repelled the dye is reached. In some cases, there may be a period of several days to allow the wax to build up before spraying is safe to proceed, particularly after a period of poor weather conditions.

In peas, the main post-emergence herbicide currently used in Europe is bentazone, sometimes in mixture with the phenoxybutyric pesticide MCPB, and it is important that wax level is high when using such a mixture. For *Vicia* and *Phaseolus* beans, bentazone is the only material safe to use as a post-emergence material for the control of broadleaved weeds.

Translocated herbicides

Here the herbicides are taken up and transported within the plant vascular system. Selectivity depends on the resistance of the pea and the susceptibility of the weed in reaction to the herbicide. The pea may have a mechanism for breaking down the herbicide to a non-toxic derivative, or conversely the susceptible weed may metabolize a toxic derivative by a biochemical process that the pea does not possess. An example of this is non-toxic MCPB, which is converted to the toxic phenoxyacetic MCPA by β-oxidation in susceptible broadleaved weeds, whereas legumes lack this metabolic process. Selective use of bentazone may be attributed to differential retention and absorption and the ability to metabolize bentazone and this material is commonly used in *Vicia* and *Phaseolus* beans.

CHEMICAL MANAGEMENT OF WEEDS

Peas

Well established pre-emergence herbicides for the control of most annual weeds in the UK are based on mixtures or tank mixes of active ingredients with pendimethalin. These include clomazone, which is especially useful for the control of cleavers (*Galium aparine*), linuron, a general-purpose residual herbicide, and imazamox, which is also useful where volunteer oilseed rape is present as well as giving some suppression of volunteer potato shoots.

The only post-emergence materials that are used are based on a tank mix of bentazone, which is effective on broadleaved weeds, and MCPB, which again is effective on small oilseed rape seedlings but also on thistles. Grass weeds such as wild oats and black-grass can be controlled with post-emergence graminicides such as fluazifop-p-butyl, cycloxydim, quizalofop-p-ethyl or teproxydim, whilst pre-emergence application of triallate is useful against wild oats (PGRO, 2016).

Faba beans

In the UK and the rest of Europe, it is essential that pre-emergence residual herbicides are used since there is only one post-emergence herbicide available. There are no herbicides to control thistles and docks (*Rumex* spp.) as translocated materials such as clopyralid or MCPB and MCPA are damaging to beans. Most pre-emergence products have a minimum planting depth requirement and dose rate may be influenced by soil type. Where cleavers (*G. aparine*) are expected to be a problem, clomazone in a tank mix is effective. Imazamox and pendimethalin or prosulfocarb can be used pre-emergence in both spring and winter beans but linuron can only be used in the spring crop. In winter beans, limited broadleaved weed control is provided by propyzamide or carbetamide but these are mainly used for grass weed control, especially where black-grass populations are resistant to other graminicides. Bentazone is the only post-emergence material for broadleaved weed control. Triallate is often used for wild oat control as are also the quizalofop materials or tepraloxydim (PGRO, 2016).

Orobanche in Faba beans

Bean broomrape (*Orobanche crenata*) is an obligate parasite that can occur on several legumes and some Umbelliferae in cultivated fields and occasionally in gardens (Fig. 5.1). It is native to the Mediterranean region, south-west Asia and eastwards to Iran, and along with its related species *O. aegyptiaca* is regarded as a serious problem in Egypt, Jordan, Tunisia, Lebanon and Italy; it is also common in Morocco and there have been reports of serious problems in Spain, Turkey and Malta. Estimates of loss in infested beans vary from 5% in Spain to 20–60% in Morocco. In several countries *O. crenata* is a notifiable pest for which there are strict import controls. It has also occurred recently in the UK (PGRO, 2015b).

O. crenata produces usually unbranched flowering stems up to 100 cm tall that lack chlorophyll. The base of the stem below ground is normally swollen and tuberous. The inflorescence occupying up to half the length of the stem

Fig. 5.1. *Orobanche* in faba beans.

carries many orchid-like flowers arranged in spikes. Each flower can contain a capsule that may contain hundreds of seeds. *O. crenata* establishes a connection to a host root within a few days of germination, stimulated by root exudates of the host. Seed needs to condition in a moist environment and suitable temperature for several days prior to germination. Otherwise mature seeds can remain viable in a dormant state for many years.

As *Orobanche* spp. are totally dependent on their hosts for nutrition, the effects on the host are proportional to the biomass of the parasite. Symptoms of the presence of *O. crenata* may not be apparent until after emergence of the parasite; a reduction in pod setting and seed development is certain and may be serious. The beans may wilt and, in severe cases, collapse.

There is no single effective management technique. The main approaches involve phytosanitary measure to control seed and plant movement. Cultural control techniques should be used in combination, as no single method will have the maximum effect. Catch cropping can be employed whereby the host plant is sown and then ploughed in before the parasite has reproduced. Some effect from the use of soil sterilants such as dazomet has been reported to give some degree of control. Breeding efforts are currently in progress to develop resistant *Vicia* varieties (Rubiales, 2014).

Phaseolus beans

In the UK and Europe, there is a limited range of herbicides available for this crop. The contact-acting herbicide diquat applied before the crop emerges will kill weeds present at the time of spraying. Pre-emergence use of S-metolachlor, linuron and pendimethalin plus clomazone is permitted and can provide useful pre-emergence weed control. However, there is date restriction on the use of S-metolachlor. Bentazone is the only post-emergence herbicide available for broadleaved weed control. Where weed problems are severe, an adjuvant oil may be added but crop injury effects may be increased. Wild oats can usually be controlled by cultivation prior to sowing. Cycloxidim controls annual and also perennial grasses but not annual meadow-grass (*Poa annua*).

SUMMARY

Weeds are a problem in all crops as most can be competitive to the crop when present in high populations, or can be contaminants as seed heads, flowers or stem fragments in vegetable crops that are harvested mechanically. The cost of weed infestation to productivity is high. The armoury of chemical aids to control weeds in peas and beans is very small and the number of active ingredients for effective selective weed control is not likely to increase. This is now leading to an increased reliance on cultural methods of weed reduction and increasing innovations in management techniques.

REFERENCES

Burnside, O.C., Wiess, M.J., Holder, B.J., Ristan, E.A., Johnson, M.M. and Cameron, J.H. (1998) Critical periods for weed control in dry beans (*Phaseolus vulgaris*). *Weed Science* 46(3), 301–306.

Davies, D.H. and Welsh, J.P. (2002) Weed control in organic cereals and pulses. In: Younie, D., Taylor, B.R., Welch, J.M. and Wilkinson, J.M. (eds) *Organic Cereals and Pulses*. Papers presented at conferences held at the Heriot-Watt University, Edinburgh, and at Cranfield University Silsoe Campus, Bedfordshire, 6 and 9 November 2001. Chalcombe Publications, Southampton, pp. 77–114.

Grundy, A.C., Mead, A., Bond, W., Clark, G. and Burston, S. (2011) The impact of herbicide management on long-term changes in the diversity and species composition of weed populations. *Weed Research* 51(2), 187–200.

Haramoto, E.R. and Gallandt, E.R. (2005a) Brassica cover cropping. I. Effects on weed and crop establishment. *Weed Science* 53, 695–700.

Haramoto, E.R. and Gallandt, E.R. (2005b) Brassica cover cropping. II. Effects of growth and interference of green bean (*Phaseolus vulgaris*) and redroot pigweed (*Amaranthus retroflexus*). *Weed Science* 53, 702–708.

Knott, C.M. (2002) Weed control in other arable and field vegetable crops. In: Naylor, R.E. (ed.) *Weed Management Handbook*, 9th edition. British Crop Protection Council. Blackwell Science Ltd, Oxford, UK.

Knott, C.M. and Halila, H.M. (1988) Weeds in food legumes: problems, effects and control. In: Summerfield, R.J. (ed.) *World Crops: Cool Season Food Legumes*. Kluwer Academic Press, Dordrecht, the Netherlands.

Knott, C.M., Biddle, A.J. and McKeown, B.M. (1994) *PGRO Field Bean Handbook*. Processors and Growers Research Organisation, Peterborough, UK.

PGRO (2015a) *Notes on Growing Organic Pulses*. Technical Update 31. Processors and Growers Research Organisation, Peterborough, UK.

PGRO (2015b) *Bean Broomrape*. Technical Update 11. Processors and Growers Research Organisation, Peterborough, UK.

PGRO (2016) *PGRO Pulse Agronomy Guide 2017*. Processors and Growers Research Organisation, Peterborough UK.

Rubiales, D. (2014) Legume breeding for broomrape resistance. *Czech Journal of Genetics and Plant Breeding* 50(2), 144–150.

Tollsten, L. and Bergstrom, G. (1988) Headspace volatiles of whole plant and macerated plant parts of *Brassica* and *Sinapis*. *Phytochemistry* 27(12), 4013–4018.

Turk, M.A., Lee, K.D. and Tawaha, A.M. (2005) Inhibitory effects of aqueous extracts of black mustard on germination and growth of radish. *Research Journal of Agricultural and Biological Science* 1(3), 227–231.

Woolley, J. and Davis, J.H.C. (1991) Agronomy of intercropping with beans. In: van Schoonhoven, A. and Voysest, O. (eds) *Common Beans – Research for Crop Improvement*. CAB International, Wallingford, UK, pp. 723–725.

6

MANAGEMENT OF PESTS AND DISEASES

Pests and diseases of peas and beans are capable of reducing yields, spoiling quality, jeopardizing the reliability of production and disrupting throughput at the packing and processing facilities. Very occasionally there are dramatic losses, due to infections of disease at epidemic proportions, or due to unusually heavy infestations of a particular pest. Both are often linked with climatic conditions. There are less serious losses, such as where patches of diseased plants appear, or when pest infestation is not heavy and there are subtle and often unnoticed losses that may result from a gradual build-up of a soil-borne pest or pathogen, but gradually reducing vitality and profitability.

Control methods are now becoming more reliant on management and measures of prevention rather than a direct approach of treating when symptoms or pests appear. The reliance on synthetic pesticides is being discouraged and more emphasis is being made on prediction, forecasting and monitoring as a means of providing an avoidance strategy or a managed approach, by identifying the optimum timing for the application of pesticides and justification for their use. In many countries that produce processed crop products, there is a strong emphasis on the need for crop traceability from the field to the factory, where every input – agronomic, crop nutrition and pesticide application – is recorded by the producer and the records remain available for inspection for some time after harvest. The UK Assured Produce Scheme known as Red Tractor has been in operation for several years as a voluntary standard set up by the food industry and a similar scheme is used in Europe (*GLOBALG.A.P.*). Both schemes include crop protocols that are standardized in agreement with retailers, food processors and merchants to provide a means of transparency and food safety for the consumer.

Many broad-spectrum pesticides were introduced in previous decades that have since been found to have negative effects on the environment and on non-target organisms. The fate of these older pesticides in the soil or water has been linked to a build-up of residues that take many years to degrade.

Recent legislation has resulted in the withdrawal of many of those active ingredients that pose risks to the environment or the operator and the introduction of new-generation pesticides can now only be made possible after stringent testing by the manufacturers. Even then, new research can often find unexpected problems that may be associated with the pesticide or its by-products.

The use of pesticides is now becoming less, especially in Europe and increasingly elsewhere, as pressure from retailers demands quality assurance from the producers of crops for human consumption that involve full transparency and of the justification for all measures, including pesticide usage on the crop.

Management of pest and disease control has therefore moved more towards an integrated approach utilizing whatever methods of prediction, forecasting or monitoring that are practical before making decisions on the use, or not, of pesticides. This involves all aspects of husbandry from the use of healthy seed, the choice of field, variety and the cropping rotation, frequent crop monitoring with monitoring systems where they are available and detailed understanding of the crop and the biology of the pest or pathogen.

Peas and beans (*Vicia* and *Phaseolus* species) can be affected by a wide variety of pests and diseases and as the crops are grown in many areas of the world it is not possible to describe each one in this chapter. However, some of those that have major significant effects are included for the three crop species. More complete listings together with illustrations of the pests and their damage and symptoms of diseases are available in several other publications (Allen and Lennie, 1985; Kraft and Pfleger, 2001; Schwartz *et al.*, 2005; Biddle and Cattlin, 2007).

As all the described crop species are legumes, they can be hosts to the same organism even when there are slight differences in the races or strains, particularly with fungal pathogens, and so where there are commonalities these are identified. In each case, a description of the symptoms, effect on the crop, biology of the pest or disease and methods of monitoring and identifying the economic thresholds for justification of treatment are described. However, specific pesticides are not included as the availability of these may change and different pesticide products are not always available in every country or state.

PESTS OF PEAS AND BEANS

Peas

Pea and bean weevil
The crops most commonly affected by the pea and bean weevil (*Sitona lineatus*) are peas and *Vicia* beans. The pest is present in most temperate areas of the world where legumes are grown, though frequently most damage is seen in Europe. The foliage of seedlings and young plants affected by the weevil shows

characteristic feeding damage in the form of semi-circular notches around the edges of the leaflets. This may be slight or severe and severe damage can retard growth, particularly during periods of slow growth when temperatures are low. In severe instances the yield of peas and beans can be reduced by about 25% but often crops will grow away from low damage levels; however, larvae developing from eggs laid during feeding will feed below ground on the root nodules. This can reduce the nitrogen availability to the plant and result in nitrogen deficiency.

In the spring, adult weevils that have migrated from field margins where they overwinter begin feeding on newly emerging seedlings, mate and lay eggs at the base of the plants. The eggs are then washed down by rain and after a few days the larvae hatch and make their way to the nitrogen-fixing nodules on the roots. After pupation, the newly emerged adults feed on green tissue before returning to the overwintering sites.

Control of the pest is very difficult, as the adults are active during the day when temperatures are relatively high but when seedlings are very small. The insects favour cloddy dry soil conditions and targeting the pest with insecticide sprays is often hit or miss. To maximize the effects of control, the pest needs to be controlled at a time of feeding but before the adult has laid eggs. A monitoring system designed to assist in timing the sprays and to assess the likely level of damage based on insect numbers was developed in the UK, using a lure containing a synthesized combination of insect aggregation pheromone and legume leaf volatiles. This is deployed in a funnel trap, placed in fields at the time of drilling. Sprays can then be applied if a threshold catch has been recorded in the trap (Biddle *et al.*, 1996).

S. lineatus is also host to a pathogenic fungus, *Beauvaria bassiana* (Steenberg and Ravn, 1996; Maurer *et al.*, 1997), the spores of which germinate on the surface of the insect and penetrate the cuticle following germination. The fungal hyphae then invade the body tissues and eventually the insect dies. Current studies are examining the use of the aggregation pheromone as an attractant to lure the insects to a bait containing *B. bassiana* spores (Bruce, 2016).

Pea aphid

Of all the pests of peas, the pea aphid (*Acyrthosiphon pisum*) is the most common and the most damaging, both as a direct feeder and as a vector of yield- and quality-reducing plant viruses. It is found in many temperate countries, including North and South America, Europe, South Australia, Tasmania and New Zealand. It lives on a wide range of leguminous plants, including peas, clover, vetch, sainfoin, faba bean and broom (Fig. 6.1).

Infested peas are retarded, the tops become chlorotic and leaves puckered, and the excretion of honeydew by the aphids encourages colonization by secondary moulds such as *Cladosporium* spp. or *Botrytis cinerea*. Leaves and pods may be distorted, coupled with underdeveloped peas. In addition, aphids transmit pea enation mosaic virus (PEMV), pea top yellows virus (PTYV) and pea

Fig. 6.1. Pea aphid.

seed-borne mosaic virus (PSbMV), all of which are damaging to yield and quality.

Although other aphid species are found on peas, *A. pisum* is the most common and damaging. Most crops become infested at some time during the growing season. Winged migrants move into crops in the early summer and colonies develop rapidly around the growing points and beneath the leaves. Reproduction and colony build-up is determined by the health of the crop and temperature. Where temperatures exceed 23°C, aphid reproduction slows down (Morgan *et al.*, 2001). Colonies also develop on pods and feeding continues until the pods mature, when a winged generation occurs that migrates to overwintering hosts such as clover and lucerne, where eggs are laid. Adults are able to survive mild winters and infest peas early the following year. At this stage, the peas are especially susceptible to virus infection and viruses such as pea top yellows virus can become serious.

There have been various estimates made of the economically important sizes of populations infesting vining peas (Biddle *et al.*, 1994). In Europe and elsewhere, it is common practice to treat vining peas as soon as infestations are noticed. In the UK, it has been shown that significant yield increases in combining peas can be obtained from using aphid control up to the time at which plants have produced four pod-bearing nodes, after which no economic benefits are obtained.

Hoverflies (Syrphidae) are important predators of pea aphid but on occasions contamination of vined peas by hoverfly larvae and puparia have led to the rejection of crops destined for freezing (PGRO, 1970). In combining peas, loss of yield remains the main problem. Whilst aphids have not developed resistance to the commonly used aphicide, pirimicarb, there is concern that repeated applications of a single active ingredient will lead to resistance developing.

Pea seed beetle

The pea seed beetle (*Bruchus pisorum*) is found in the USA, Australia and many warmer European countries, but not in the UK and Scandinavia. Peas grown for dry harvest and destined for premium food markets or for seed are blemished by the holes left by the emerging adults. Locally high populations of bruchids can develop and once established in an area the infestation level can be too high for the pea processors to clean and sort. Where mechanical cleaning is available, such cleaning incurs price penalties to the producer. Damaged seed may still germinate but smaller seed can succumb to pre-emergence mortality in the field and damaged pea seed is considered undesirable by the end user. The presence of live beetles in seed lots can result in the seed lot being classed as uncertifiable and it may have to be fumigated (Fig. 6.2).

Adult beetles leave their overwintering site in early summer when maximum temperatures reach 20°C. They fly to flowering crops and lay eggs on newly developing pods. After hatching, the larvae bore through the pod wall and into the developing seed, where they feed on the cotyledon until mature. As the seed matures, the larvae pupate and emerge as adults by cutting their way out of the seed, leaving the characteristic circular hole.

Control is difficult in the field, as the beetles feed within the pea canopy and are not often exposed to insecticides. Early-flowering varieties may escape damage if the adult flights from the overwintering sites are delayed by cool temperatures. Work is in progress to identify resistance characteristics in peas based on work carried out in Australia on *Pisum fulvum*, but transgenic resistance development is not commercially practised on peas at the present time.

Fig. 6.2. Adult pea beetle emerging from seed.

Work in Chile has identified a possible means of biological control with the parasite *Uscan senex* (Pintureau *et al.*, 1999) and other parasitoids have also been described but no commercial development has been introduced.

Pea midge (pea gall midge)

The distribution of the pea midge (*Contarinia pisi*) is restricted to the more northern temperate areas of Europe, Scandinavia and the UK. Vining peas are the most severely affected crop, as pea varieties have been developed to produce short flowering periods to allow a more even maturity and this results in a large proportion of flower buds susceptible to attack by the midge larvae. Older pea varieties and most combining peas are more indeterminate and have a longer flowering period and suffer less damage.

Crop damage is caused by larval feeding within the flower bud, which results in a distortion of the bud and a 'nettle head' caused by foreshortening of the flower stalks. Damaged buds fail to produce pods. Crops are susceptible to damage when they have reached the enclosed bud stage, where midges (if present) lay eggs on the outside of the bud, allowing larvae access to the developing bud, where they feed within the base of the flower. Yield loss can be as much as 50% where there is a larger number of flower buds within the apical shoot damaged by the larvae.

Adult midges emerge from the soil of the previous year's pea crop in late spring and mate before the females begin to migrate to the new season's crop. The male midges remain in the field for a short while before dying. Females fly to nearby crops, where eggs are laid within the flower bud clusters. After feeding, the larvae fall to the soil, where they move down the soil cracks to overwinter as cocoons. In late spring the larvae break out of the cocoons and pupate before emerging in early summer.

Because the midge has a very short flight range, cultural control can be effected by growing peas away from previously infested sites. This may involve moving areas of production for 1–2 years to break the cycle but it has been successful in Sweden (Jönsson, 1988). Chemical control is by insecticide spraying as soon as midges are present in the crop but timing is best achieved by deploying a monitoring system, which is a pheromone-based sticky trap that detects the early migrating midges (Hillbur *et al.*, 2001; PGRO, 2015).

Pea moth

Pea moth (*Cydia nigricana*) is particularly troublesome in areas of intensive combining pea production and where vining peas or fresh market peas are grown in the same vicinity, allowing large populations of the pest to develop. It is found throughout the temperate areas of Europe and some localized populations occur in some western states of the USA, southern Canada and Japan.

Damage is caused by the larvae feeding within the pods on the developing seed. The damaged peas have irregular-shaped or circular holes of varying sizes and the pods contain frass that is webbed together. Produce is therefore spoiled and damaged seed is likely to fail to germinate. In combining peas, damaged seeds have to be removed by a costly cleaning operation (see Chapter 7) but in vining or fresh market peas, crops are likely to be rejected by the processor or retailer (Fig. 6.3).

Adults emerge from the soil of the previous year's pea fields and fly to flowering pea crops in early summer. Eggs, laid on the foliage and petals, hatch after about 10 days and larvae find and bore into developing pods and begin feeding on the peas within. When mature, the larvae leave the pods and fall to the soil, where they bury themselves before producing an overwintering cocoon.

Because moth populations build up locally where peas have grown to maturity, any peas growing within the location the following year are susceptible to infestation. Chemical treatment is the only effective means of control but sprays have to be applied as the larvae hatch from the eggs and before they burrow into the developing pods. An effective monitoring system was developed in the UK and is used commercially to ascertain the time of moth activity and a prediction of the optimum time for spraying.

The monitoring system is based on the use of a synthetic analogue of the female pea moth sex-attractant in sticky traps, which are put into the pea crop just before flowering. By monitoring trap catches of male moths and daily temperatures, the user of the system identifies when a threshold catch has been recorded and can compute the correct spray date (Biddle *et al.*, 1983).

Fig. 6.3. Pea moth larva.

Pea cyst nematode

Known also as pea root eelworm, the pea cyst nematode (*Heterodera gottingiana*) is a soil-borne pest that can infest peas, faba beans, vetches (*Vicia sativa*), lupins (*Lupinus* spp.) and sweet peas (*Lathyrus oderatus*). The pest has been recorded in Europe, Russia and the USA and there have also been reports of the pest occurring in Japan, China and the Middle East. The most seriously affected crop is peas; *Vicia* beans can also become infested but it is rare that serious symptoms or effects are noticed. Infestation occurs in discrete patches of soil where the nematodes multiply and form cysts, which remain viable in soil for many years or until a host crop is grown. Affected peas become stunted and pale and there is an absence of root nodules. The roots that have been invaded by the nematode contain the swollen bodies of the female nematodes, which later erupt from the root tissue and dry, forming numerous brown-coloured lemon-shaped cysts. The plants eventually senesce and die and fail to produce harvestable pods. Yield loss can be total in infested areas of the field; and in surrounding areas where plants are showing less severe symptoms, they may mature prematurely. *Vicia* beans are more often unaffected by an infestation, though occasionally affected areas in an infested crop can remain stunted.

Once the cysts have been produced, the length of viability can be up to 20 years and the only means of control is by not growing a legume host in the field for at least that period of time. Cysts can also be transported to other fields in soil on implements, and it is thought that the pest has been introduced to other areas by planting imported bulb crops that have been produced in old infested pea fields.

In practice, no chemical means of control are used. Fields where infestations have been recorded in the past are not cropped with peas or *V. faba* for at least 20 years. Soil sampling and extraction of cysts can be carried out but often small infested areas can be missed during field sampling.

Faba and broad beans

Stem nematode

Ditylenchus gigas, formally known as the 'giant' race of *Ditylenchus dipsaci*, is a free-living nematode that can survive in soil for several years. It has been found affecting *Vicia* beans in many areas of the world, especially where *V. faba* has been grown for many years. The species is specific to *Vicia* beans and although other species of nematode (*D. dipsaci*) may infest beans, they are not a cause of serious yield loss. Damage is often observed as plants reach flowering, but young seedlings may develop severe symptoms before this stage, particularly in wet conditions. The plants are stunted and the stems thickened and twisted. There is often blistering of the stem and this may develop a reddish colour. Infested stem can collapse. Leaves become distorted at the petiole and the growing points may be affected in a similar way. Pods fail to develop and where seeds are produced they are black and shrivelled, or the testa may be discoloured in small patches. Affected plants occur singly or in large areas of the field and both winter- and spring-sown beans can be affected.

The effect on yield is severe where high populations of the pest are present. In broad beans for human consumption, seed may be discoloured and their presence can lead to rejection by processors.

Stem nematodes are both seed borne and soil borne. Once infested, the nematodes move within the plant and multiply in the tissue before a proportion of them move into the developing seed and mass beneath the testa. They can then withstand desiccation and survive within the seed for up to 2 years. Infested seed can be transported and, when planted, the nematodes re-infest the developing plant and populations move to surrounding plants, thereby starting a new infestation in otherwise clean fields.

Because the pest can be seed borne, the only means of control is by producing seed in non-infested fields and by stringent testing of the seed. The method of testing is based on extracting nematodes from seed by soaking in water for 24 h and examining the soak water microscopically (Augustin and Sikora, 1989) and such a method is used by many seed testing laboratories. Currently an evaluation of molecular testing procedures is being carried out by several European researchers. Where a sample is found to contain nematodes, the seed should not be used for sowing (PGRO, 2015). There have been attempts at screening *V. faba* lines from the ICARDA germplasm collection for resistance to *D. gigas* and breeding work is at an advanced stage of development in producing a variety of field bean with some resistance to the nematode (Stawniak, 2011).

Black bean aphid

Aphis fabae is mainly a pest of faba beans and occasionally of peas, but may be found in high numbers on *Phaseolus* beans, spinach and sugar beet where populations that colonize the underside of the leaves can become chlorotic and crinkled. *A. fabae* is present in many countries, including Scandinavia, Europe, Asia, the Middle East, parts of Africa and North and South America, but it is not common in the tropics. The related black aphid *A. craccivora* is common in warmer countries where temperatures around 24°C are more favourable for its development. This species has a wide host range and includes most leguminous crops and plants (Fig. 6.4).

Colonies of *A. fabae* build up rapidly on bean plants particularly towards the growing shoot, where they cause direct feeding damage that reduces flowering pod set and apical growth. As well as feeding, they are vectors of viruses including bean leaf roll virus (BLRV) and bean yellow mosaic virus (BYMV). In association with the pea aphid, they can also transmit pea enation mosaic virus (PEMV) which, in combination with BLRV, can cause significant reduction of yield. Yield reduction varies as to the degree of infestation but yield loss is high where colonies are dense around the flowering and pod-bearing nodes.

Fig. 6.4. Black bean aphid.

The effect on yield and quality of virus infection also can vary from slight to severe with distortion and poor seed development or staining of the seed coat, leading to crop rejection.

Aphids migrate to bean crops in late spring or early summer and colonize the tops of the plants before developing as large dense colonies of many hundreds of individuals.

After colonization, winged aphids leave the crop and lay eggs on overwintering hosts, including the spindle (*Euonymus europaeus*). *A. craccivora* has a similar life cycle to *A. fabae* but the woody overwintering hosts are laburnum (*Laburnum anagyroides*) and broom (*Sarothamnus scoparia*).

Control of both species is based on insecticide treatment applied when colonies are first observed and before flowering, as early migrating aphids can begin virus transmission as soon as feeding commences.

Bean seed beetle

As one of the many species of bruchid beetles, the bean seed beetle (*Bruchus rufimanus*) is the most damaging pest of *V. faba* and is present in many temperate countries.

The main effect on the beans is the infestation of the developing seed by the larvae and pupae of the insect (Fig. 6.5). Fresh beans may be blemished by the larval entry holes and these become apparent when the broad bean pods have been opened. Beans will also contain larvae at this early stage and

Fig. 6.5.   Bean seed beetle.

produce for processing or fresh market is unsaleable. In the dried crop, the presence of pupae or un-emerged adults causes problems with the export of beans for human consumption and beans with exit holes are also unsuitable for high-quality markets unless they are removed mechanically during the cleaning operation.

Unlike closely related bruchid beetles, *B. rufimanus* has only one generation per year and relies on its development on the growing crop. It is not a pest of stored produce. Adults fly in to beans at the onset of flowering and feed on pollen for up to 2 weeks before they become sexually mature. At that point, females lay eggs on developing pods; after a few days the eggs hatch and the larvae burrow into the pod and into the developing seed. Feeding commences on the cotyledon until both the seed and the larvae have reached maturity. At this stage the larvae pupate and emerge as adults just before or just after harvest, leaving circular exit holes in the bean. Adults then fly to their overwintering sites, which include woody hedgerows and under tree bark in scrubby vegetation surrounding the field.

Attempts at providing a monitoring system for the presence of bruchids in the crop have been carried out and there has been some limited success using cone traps with lures of flower volatiles, which have been shown to be effective in trapping adults as they move to crops but not attractive enough to compete with the crop once flowering has commenced (Ward and Smart, 2011). Decision-support systems for bean growers to time spray applications are in operation in France and a similar but web-based system (Bruchidcast) has been developed recently in the UK. Both systems utilize day temperatures to ascertain the activity period of bruchids so that insecticides can be applied to coincide with oviposition.

Some limited screening for resistance to *B. rufimanus* in *Vicia* beans has been carried out but as yet no commercial development of resistant varieties has been achieved.

Phaseolus beans

Bean seed fly (seedcorn maggot)

Bean seed fly (*Delia platura*) is a very common pest and can be found in most temperate countries affecting a wide range of large-seeded crops, including peas, beans, marrows, lupins, maize and soya beans. In severe infestations large numbers of seedlings can be lost, severely affecting plant populations. Seed of late-sown peas or beans is attacked by the larvae of the fly during germination. Faba beans are seldom damaged as they are not frequently sown during the early part of the summer, but *Phaseolus* beans and late-sown peas for freezing or fresh market are particularly susceptible.

Larvae from eggs laid in freshly disturbed soil feed on decaying vegetable matter and also tunnel into freshly imbibed seed. The larvae feed within the cotyledons and damage the developing plumule and root. Tunnelling can take place within the stem and the growing point can be damaged resulting in a 'baldhead' or 'snake head' symptom, where the stem elongates but there is an absence of terminal leaves. Secondary shoots can develop from the seed in peas and *Vicia* beans, but in *Phaseolus* the seedlings fail to compensate for the damage. Severely infested seeds fail to produce a seedling and decay before emergence. Damage is often noted in patches in a field, as the flies tend to aggregate before oviposition.

In late spring, adults congregate over freshly disturbed soil in fields that contain large amounts of green weed vegetation or fresh crop debris. Late-cultivated soils where weed infestation has developed are more likely to be infested by *D. platura*. Flies lay eggs on the soil surface where the larvae are able to feed on the vegetable debris but also tunnel into large seeds immediately after imbibition has occurred.

To reduce the risk of infestation, seedbed preparation should ensure the burial of any weed growth, particularly if the weeds have not desiccated, and soils with a high level of crop residue such as carrots, sugar beet or bulb crops should be avoided. In small-scale production, peas or beans can be protected with fleece or plastic film until germination has occurred, but in commercial crops of *Phaseolus* beans an insecticide seed treatment is often used.

Mexican bean beetle

Mexican bean beetle (*Epilancha varivestis*) is perhaps the most serious insect pest of dry beans in the western areas of the USA. It is believed to be native to the plateau region of southern Mexico but the insect is found in the USA (in most states east of the Rocky Mountains) as well as in Mexico. In eastern regions of USA, the pest is present wherever beans are grown, while western infestations are in isolated areas, depending upon the local environment and precipitation. The insect is not a serious pest in Guatemala and Mexico but is very abundant in several areas in the western USA. The southern limit of the known distribution is in Guatemala and the northern limit is southern Canada and New England.

Mexican bean beetle adults overwinter in debris in fields, along field margins. The beetles move out into the bean fields in June and July and females begin to lay egg masses on the beans after they have fed for 1–2 weeks. Larvae and adults feed on the underside of leaves, stripping the epidermis from the leaf and leaving a skeletonized leaf. Management of the pest involves monitoring and searching for egg masses and this allows the grower to make a timely decision on the need to treat before damage has begun (Sanchez-Arroyo, 2015).

MAJOR DISEASES OF PEAS AND BEANS

Peas

Foot rot diseases

The most damaging complex of soil-borne root-infecting pathogens contains several fungi that can cause severe basal foot rot or root rot in peas in virtually all of the traditional pea-growing areas of the world. The effects on the crop have been described as pea foot rot (Biddle, 1983). The fungi build up in the soil as a result of frequent cropping and because the decay rate of the resting spores is slower than the return time of the pea crop, the soil-borne population increases to the point at which the crop becomes heavily infected. In the UK, three principal organisms are involved in the foot rot complex, each of which can infect individually or in concert. Several *Fusarium* spp. can be present, the principal being *Fusarium solani* f. sp. *pisi*. On its own, the pathogen invades the roots early in the growing season; during its progression, the root decays and the vascular system becomes blocked with toxins produced by the fungus. This stunts the plant and causes premature senescence and subsequent death. The second pathogen is *Didymella pinodella* (syn. *Phoma medicaginis* var. *pinodella*) (Chen and Cai, 2015), an ascomycete fungus that can be soil borne, through the production of thick-walled chlamydospores, or seed borne once the infection develops on the foliage and pods. Initially *D. pinodella* infection causes a girdling of the base of the stem, which becomes blackened and eventually collapses. Leaf and pod spotting can occur after the initial infection on neighbouring plants and the pods and seeds are infected during periods of wet and rainy weather. The third pathogen is *Aphanomyces euteiches*, which causes a root rot. This is a phycomycete fungus that produces zoospores, which are motile in wet soil conditions. Once infected, the root cortex decays and sloughs off the vascular strand. Oospores produced in the cortex can then remain in the soil for many years. It is favoured by wet soil conditions and is a principal pathogen in the lighter finer structured soils in the Paris basin in France, where large-scale losses of peas to root rot were often experienced when peas were grown frequently on farms.

There is no control of these soil-borne fungi and the only effective means of management is to avoid cropping land with peas more frequently than once in 5 years (Biddle, 1983). In the UK, where problems developed severely in the 1970s, the cropping interval for vining peas has been increased to 6 or even 7 years to prevent damaging populations of the fungi from developing.

In the UK, a soil-borne disease prediction test was developed and is used as a guide to detecting fields with a high population of the pathogens (Biddle *et al.*, 1988). In France and Scandinavia a similar method is used to pick out fields with high levels of *A. euteiches*. Much work is in progress to breed varieties with resistance to these pathogens but because of the complexity, no one variety has been introduced with combined resistance to all (Ali *et al.*, 1994), although current work in Europe with a cooperative of research is in progress to identify resistance to *Fusarium* and drought stresses.

Leaf and pod spot (*Ascochyta*)

In peas, there are three closely related fungi that comprise the ascochyta blight disease complex: *Ascochyta pisi*, *Mycosphaerella pinodes* and *Didymella pinodella* (*Phoma medicaginis* var. *pinodella*) and all can cause leaf and pod spot in peas. All are seed borne and can survive on crop debris; *M. pinodes* and *D. pinodella* are also soil borne. Symptoms of each are different, but *A. pisi* produces the more distinctive type of leaf and pod spot and is more commonly found worldwide (Fig. 6.6).

Seedlings derived from seeds infected with *A. pisi* show symptoms shortly after emergence, particularly if the weather has been wet. Brown to grey sunken lesions develop on the stem, leaves or stipules. As the plant develops, pycnidia form in the centre of lesions and ascospore release is triggered by rain or water droplet splash. The spores then infect surrounding plants, eventually producing lesions on the pods and deep-seated infection continues to infect the developing seed. *M. pinodes* again arises from infected seed but can also infect from soil-borne crop debris or from chlamydospores that infect the stem base, resulting in stem girdling and subsequent leaf and pod spotting with smaller, more purple-coloured lesions, which are not as deep seated as *A. pisi*. *D. pinodella* is more likely to infect from the soil, where thick-walled chlamydospores can remain viable for several years. The disease primarily causes a foot rot, but leaf

Fig. 6.6. *Ascochyta pisi.*

and pod spotting can occur later in the season if soil conditions are wet with frequent rain and warm temperatures.

The main source of primary infection is the seed and the use of healthy seed is the only way to avoid *A. pisi*. Because both *M. pinodes* and *D. pinodella* are soil borne, a long rotation where peas are not grown too frequently will delay the build-up of soil-borne populations. Seed crops produced in arid areas are less likely to become infected but in other temperate areas disease risk is higher where summer rainfall is more erratic. Fungicidal seed treatments are effective in controlling most seed-borne *Ascochyta* but the use of highly infected seed, even with seed treatment, does not always produce a complete stand of healthy seedlings.

Resistance to *A. pisi* is currently being investigated but because of the range of pathotypes present it is difficult to produce varieties with complete resistance (Tivoli *et al.*, 2006; Muehlbauer and Chen, 2007).

Powdery mildew

The disease caused by *Erysiphe pisi* is favoured by warm temperatures during the day followed by cool, humid nights. Late-sown peas are more susceptible to infection and it is in these crops that the effects are greatest. Symptoms appear as small irregular patches of powdery mildew on the leaf surface of the lowest leaves and then there is rapid progression to the rest of the plant and pods. It is a very widely distributed pathogen in all areas where peas are grown. Severe infection reduces pod weight and yield, and seeds fail to develop their full potential in infected pods. The maturity of peas grown for dry harvest is delayed and peas grown for the fresh market are blemished and unsaleable. The fungus overwinters on crop debris and alternative host plants such as vetches and other wild legumes.

Varietal resistance to powdery mildew is largely conferred by three recessive genes, *er-1*, *er-2* and *er-3*, with *er-1* being the most frequently used by breeders (Fondevilla and Rubiales, 2012). Several varieties that are commercially available are completely resistant to the disease under field conditions and more work by breeders is continuing to extend the number. A complete list of varieties available for the fresh market and for processing with good agronomic characteristics grown in the UK is shown in Table 3.1 (see Chapter 3).

There are very few fungicides available to control powdery mildew but regular sprays of elemental sulphur can provide some protection.

Pea downy mildew

Downy mildew caused by *Peronospora viciae* is common throughout the temperate pea-growing countries, particularly in northern Europe, Scandinavia, New Zealand and the north and mid-west states of the USA, though infection is uncommon in the drier states. It causes significant plant loss early in the season when weather conditions are more favourable for infection, as it prefers humid, cool conditions. Later infection debilitates the plant and reduces pod and seed set. Pods are blemished and become unsaleable for the fresh market and peas for processing are often blemished and undersize (Fig. 6.7). Oospores are produced within infected tissue and once returned to the soil they remain viable for many years. Close cropping of peas is the main cause of severe infection.

Taylor *et al.* (1989) reported that there were 11 races or pathotypes of *P. viciae* in Britain and most fields with a history of disease contain mixtures of these. Breeding for complete resistance has therefore not been achieved but several varieties show differing levels of tolerance in the field to infection and on this basis varieties in the UK are classified for their relative tolerance. For vining peas, very few varieties show good tolerance and all are routinely treated with a fungicidal seed treatment. Before the introduction of the acyl-aniline fungicides, which are active against *Peronospora* spp., many early-sown early-maturing pea varieties were severely infected by the disease and yield loss was greatly reduced, making the cultivation of early varieties almost unviable. In contrast, most of the combining pea varieties grown in the UK and Europe have good field resistance and are grown in the absence of a seed treatment.

White mould

Sclerotinia sclerotiorum has a very wide host range and can infect most broad-leaved crops, including vegetables, potatoes, linseed (*Linum* spp.) and oilseed rape (canola) in most parts of the world. The disease spreads rapidly in warm wet weather and causes rots of the stem, foliage, pods and peas. Infection is characterized by the appearance of a wet rot rapidly followed by the production

Fig. 6.7.   Pea downy mildew.

of a dense white cottony growth of mycelium. Infected stems appear bleached and sclerotia develop inside the stems and pods. Infection is common in some areas where susceptible crops are present in the rotation and it occurs in many areas of the world except in the warm humid tropics. Once infection has become established the sclerotia are returned to the soil during harvest, where they can remain viable for 5 years or more. Under suitable conditions, the sclerotia produce apothecia on the soil surface which release ascospores that germinate on fresh green leaves, stems or moribund flower petals; the pathogen then becomes virulent and invades the rest of the plant.

Management of the disease is based on crop rotation and not growing peas in close proximity or in close rotation with previously infected hosts. Most pea varieties are susceptible, though early-sown crops may escape infection as it usually occurs later in the season when the temperature has increased. Modelling the outbreaks of infection based on petal sampling of oilseed rape and local meteorological data has been done in the UK and elsewhere for oilseed rape and similar modelling has been done on lettuce (Clarkson *et al.*, 2014). For peas, the time of susceptibility to infection is very short and models tend to indicate a wide window of opportunity for treatment. The use of a predictive model as a decision-support system for use in peas has been examined in the UK, but peas are only susceptible to infection at the flowering stage, which is very short (HDC, 2013).

Pea bacterial blight

This disease can be important in autumn-sown crops or crops grown under irrigation. In Europe and New Zealand where autumn planting in some areas is common, the disease can be very severe and crop loss is not uncommon. In dry conditions, the disease is not severe enough to cause loss; in spring-sown crops, significant damage is rare unless there has been damage by frost or hail.

The disease is caused by the seed-borne bacterium *Pseudomonas syringae* pv. *pisi* and symptoms can occur at any time during the growing season, especially following physical or mechanical damage to the foliage of the peas. Once established, under favourable conditions the infection spreads rapidly as water-soaked lesions on the stems and leaves and in particular in the stipules, where fan-shaped necrotic areas develop. Pods become pitted with grease-spot lesions and seed infection can take place during seed maturation. There are seven races of pea bacterial blight. In the UK, race 2 seems to be the most common in spring varieties, while races 4 and 6 are found in winter peas. In the USA, race 4 seems to be the common race in spring peas.

There is no chemical means of control and the use of healthy seed is the most effective means of avoiding infection. The bacterium does not survive on crop debris for more than 1 year. No varieties are resistant to all seven races but several are resistant to races 2 and 4.

Fusarium vascular wilts

Several races of *Fusarium oxysporum* f. sp. *pisi* occur in many pea-growing countries. Disease caused by *F. oxysporum* often results in a vascular wilt in large areas during the time of flowering and seed setting. Race 1 of *Fusarium* wilt is characterized by a stunting of the plant, together with a colour change affecting the whole foliage which at first turns a greyish colour before shrivelling and resulting in the death of the plant. Race 2 is known as near-wilt and plants tend to be affected either individually or in scattered areas over the field. Often the roots are decayed and the effect on the plant is again a wilt and premature senescence. Race 6 has been reported in the USA and Scandinavia and it probably occurs occasionally in Europe and Australasia. The symptoms are very similar to those caused by a soil-borne root-infecting fungus. Other races, including race 5, have been described in the USA (Kraft, 1994), but symptoms are very similar to the other races with those of race 5 being similar to those of race 1. Pea wilt became a major disease problem in the USA in the early part of the 20th century and many crops in Wisconsin were lost in most years. The disease was first described by Linford (1928) but later distinguished as race 1 of *F. oxysporum* f. sp. *pisi* in 1935. It was only when resistance to wilt was found by breeders (Wade, 1929) that the disease became manageable and now many varieties have been bred with resistance to race 1, with some also resistant to the other races.

In the UK, all registered new varieties are screened for resistance to race 1 and the information is available through the breeders or the recommended list of pea varieties (PGRO, 2015). Crop rotation is of little value once the disease is present in the soil but a rotation that includes peas no more frequently than once in 5 years will help to prevent the disease building up.

Virus diseases

Of the important pea viruses in Europe, the USA and Australasia, four aphid-transmitted viruses can cause the most serious damage or loss of crop or produce quality. The pea aphid (*Acyrthosiphon pisum*) is the vector of a number of viruses, including pea enation mosaic virus (PEMV), pea top yellows virus (PTYV), alfalfa mosaic virus (pea streak) and pea seed-borne mosaic virus (PSbMV), which is also seed borne. The most significant amounts of damage with PEMV and PTYV result from an early invasion of pea aphid bringing the diseases into the crop from overwintering hosts. PEMV has an overwintering host range that includes wild as well as cultivated legumes such as *Vicia faba*, chick pea (*Cicer arietinum*), sweet pea (*Lathyrus oderatus*), lentil (*Lens culinaris*), *Medicago arabica* and *Vicia sativa*. PTYV is also known as bean leaf roll virus and has wild hosts such as clover and vetches and in lucerne (alfalfa) and winter-sown *V. faba*. PSbMV is principally seed borne and will only transmit between nearby peas while aphids are moving into the crop. In some cases it is thought that foraging cereal aphids could also be responsible for transmitting PSbMV from an infected seedling to the rest of the surrounding crop.

Several varieties are resistant to one or both of PEMV and PTYV and a small number are also being bred for resistance to PSbMV. A possible source of resistance is a difference in the ability of the virus to pass into the developing seed (Roberts *et al.*, 2003). In general, effective aphid control is necessary to avoid infection, but in the case of PSbMV it is only the use of healthy seed that will ensure freedom from this disease. The main symptom of PSbMV is a stunting of the plant and often the produce is blemished with a white-edged marking on the testa that has been likened to the markings on a tennis ball. Affected peas can cause crop rejection by the processors of freezing peas and in the UK in the mid-1980s the seed stocks of a petits pois variety that had been widely used in the UK and Europe became seriously infected. A programme of research to identify the presence of the virus in the seed was started in 1987 and this resulted in a widespread testing programme using an ELISA-based seed test to identify the seed stocks carrying the infection. The programme was effective in allowing healthy stocks of this variety to be produced and remains one of the most important petits pois varieties grown in Europe and the USA.

In Europe, a nematode-transmitted virus, pea early browning virus (PEBV), can be found in some seasons. Free-living nematodes, including the vector *Trichodorus* sp., are frequently found in free-draining sandy soils and graze on the pea roots during periods of high soil moisture in the spring. The virus develops in peas, resulting in characteristic leaf mosaic, purple stems and stunting and necrosis of the lead shoot.

The disease can also be transmitted by seed. Seed health procedures can be effective in reducing the risk of introduction of the virus to other areas where trichodorid nematodes are present. A detailed review of the virus was written by Boulton (1996).

Diseases of *Vicia* beans

Bean downy mildew

Bean downy mildew (*Peronospora viciae*) is also found infecting peas but it has not been shown that isolates of bean downy mildew could be inoculated successfully on to pea and it is thought that there are distinct pathotypes amongst *P. viciae* populations that may be specific to *V. faba*. The disease has been reported in many temperate areas of the world but mostly where early growing conditions are cool and humid. The symptoms resemble those where peas are infected with pea downy mildew. Initial infection begins from soil-borne inoculum comprising oospores that infect germinating bean seedlings. The resulting infection develops systemically on the seedling, producing stunted pale plants on which oospores are produced from mycelium on the underside of the leaves (Fig. 6.8). Infection continues as secondary infection to surrounding plants, often ending in a complete distortion of the growing tip covered by the characteristic grey-mauve mycelium. The pathogen has not been found to be transmitted by seed.

Conditions favouring infection have been identified in a combination of air temperatures and humidity (Biddle *et al.*, 2003) which has led to a forecasting system being devised. The system that is used by growers in the UK to

Fig. 6.8. Bean downy mildew.

determine the risk of infection and a decision-support system for spray application has been made available online for UK growers (CropMonitor).

Bean downy mildew can be controlled by phenylamide fungicides and seed treatment can also be effective. There are several varieties of *V. faba* that have a greater level of field tolerance to the disease than others and this feature is provided in the UK recommended list of bean varieties (PGRO, 2015).

Chocolate spot

Chocolate spot caused by *Botrytis fabae* and also by *Botrytis cinerea* is favoured in summers featuring prolonged wet and overcast periods. It is found in all areas of the world where faba beans are grown but only causes severe losses under favourable conditions. Yield loss due to defoliation and the reduction in photosynthetic area is the main problem and beans for the fresh market that are harvested as whole pods can be blemished by the spotting and russetting that occurs on infected plants. The symptoms of chocolate spot are distinct from other leaf and pod spotting diseases in that there is no obvious production of fungal fruiting bodies such as with *Ascochyta fabae*. Autumn-sown crops are more susceptible to infection, as this can begin in the autumn and remain at low levels until the spring when weather conditions that favour development occur (Fig. 6.9). High-density planting encourages disease development and

Fig. 6.9. Aggressive chocolate spot.

date of planting will also have an influence on the amount and aggressiveness of disease that develops (see Chapter 4).

The use of fungicides to control chocolate spot is a routine procedure in most temperate countries and reliance on fungicides is total, as there are no significant differences in varietal tolerance to the disease. Timing of application is key to effective control (Gladders *et al.*, 1991).

Leaf and pod spot

Related to the pathogen that causes leaf and pod spot in peas, *Ascochyta fabae* is host specific and is the only significant pathogen to seriously affect yield and quality of *Vicia* beans. Infection can lead to collapsed plants, with heavily infected pods having deep circular or elongated lesions that affect the developing seed and a significant reduction of yield and quality. It is primarily seed borne and develops on seedlings shortly after emergence. Pycnidia produced within the disease lesions on the stems, leaves or pods contain spores that are released and splashed by rain or water drops. Internal infection can occur, resulting in deep-seated infection of the seed. Spores can also be transmitted by wind from pycnidia produced on overwintered infected crop debris where the sexual form of *A. fabae* (*Didymella fabae*) develops over the autumn and winter (Jellis and Punithalingam, 1991) (Fig. 6.10).

Chemical control has been largely unsuccessful and the health of *V. faba* beans in the UK was maintained by strict certification procedures until the mid-1990s. Since then voluntary standards have been maintained by breeders and seedsmen. Resistance to *A. fabae* was identified in winter beans in the UK several years ago (Bond and Pope, 1980) and since then efforts have been made to develop completely resistant varieties in Europe and elsewhere.

Fig. 6.10. *Ascochyta fabae.*

Brown rust

Faba beans, both spring-sown and winter-sown varieties, can become infected at any stage in the growing season by bean brown rust caused by *Uromyces fabae*. Rust is common in most bean-growing areas of the world and has become a major cause of yield loss in both field beans and broad beans.

Bean leaves become infected from flowering onwards but the most serious infections may begin later in the season, following a period of warm days and cooler nights with high humidity (Fig. 6.11). Because of the influence of weather conditions, later-maturing spring-planted beans are often more prone to infection, though in some seasons autumn-sown crops can become affected. Rust spores can survive as telia in a semi-dormant state over the winter on crop debris or on volunteer beans.

Rust causes a partial defoliation and loss of photosynthetic area, resulting in significant yield loss, especially when infection begins during the pod development phase. Pustules can develop on the stems and pods and severe infection causes poor development of the seed. In broad beans, blemished pods can make them unsaleable.

Although rust control using systemic fungicides is effective, there are some differences in susceptibility between varieties, particularly in Europe, where some varieties of spring beans are more tolerant than others. Incomplete resistance has been commonly found but hypersensitive resistance has only

Fig. 6.11. Brown rust.

recently been identified (Rojas-Molina *et al.*, 2010), though as yet no varieties have been developed with complete resistance (Adhikari *et al.*, 2012).

Virus diseases

V. faba can be infected by several bean viruses, a few of which are seed borne but most of the 18 or so viruses described worldwide are transmitted by aphids or other insects (Kumari and van Leur, 2011). Commonly experienced viruses in Europe include bean leaf roll virus (BLRV), which is transmitted by pea aphid and black bean aphid, and pea enation mosaic virus (PEMV), which is also transmitted by pea aphid. Often both diseases are present together and infected crops show various symptoms, including the classic leaf rolling, to crinkling of the upper foliage, a development of translucent spotting of the leaves, distorted and undersized pods and an interveinal chlorosis that is often confused with symptoms of manganese deficiency.

In Africa and the Middle East, one of the most common viruses is faba bean necrotic yellows virus (FBNYV), which is efficiently transmitted by *Aphis craccivora* from a range of wild host plants. In Australia, beet western mosaic virus is common in some areas and pea seed-borne mosaic virus transmitted by aphids from nearby infected pea crops can also result in quality loss due to seed coat blemishes.

As yet there are no reports of varietal resistance to any of the common viruses and attention for control and management has been targeted towards crop hygiene, weed control and pesticide application.

Diseases of *Phaseolus* beans

Aphanomyces root rot

This disease was not found to be a problem in *Phaseolus* beans until 1979, when a strain of *Aphanomyces euteiches* was identified in Wisconsin that caused severe root rot (Pfender and Hagedorn, 1982). Since then it has been found infecting crops in several states in the USA and Australia. Infection can begin early in the life of the plant or it can develop later and after field soils become waterlogged for a short period. The fungus invades the cortex of the root system, resulting in stunting and yellowing before the plant wilts and dies. Often *A. euteiches* can be found together with *Pythium* spp. as the latter pathogen is more widespread but is favoured by similar conditions. The host range of *A. euteiches* f. sp. *phaseoli* includes lucerne (alfalfa) as well as green beans and dry beans.

Some resistance has been found in breeding lines but their agronomic characteristics are too far away from commercial attributes for the near future. Heavy irrigation is associated with infection and so soil moisture content and temperature are key factors in managing the disease in large-scale production of irrigated beans. The use of green manures, especially mustard, has been shown to be useful in suppressing *A. euteiches* and other pathogens (Parke and

Rand, 1989). The incorporation of brown mustard, grown as green manure and then chopped and rolled into the soil, has a fumigant effect on the pathogen. Other work has indicated the value of other green manures, together with an integrated approach to control, using tolerant varieties and good soil management, in reducing inoculum levels in the soil (Tu, 1991).

Fusarium foot rot

Fusarium foot rot of beans occurs in most bean-growing areas throughout the world. The disease usually causes most damage when the roots are first put under stress by either soil consolidation, drought, waterlogging or oxygen stress. Early symptoms are stunting of the plant, chlorosis of the upper leaves and reddening or browning at the stem base. The roots are usually blackened and the stem base may collapse. Occasionally short runs of plants along a row may show the symptoms where soil conditions are causing stress.

Loss of plants through infection has the effect of reducing yield. Depending on the areas of infection in a field, yield loss can be severe. Where plants are short in height, the efficiency of mechanical harvesting is reduced.

Fusarium solani f. sp. *phaseoli*, the causal organism, remains in the soil in the form of chlamydospores for many years. The chlamydospores germinate in the presence of root exudates from the host and infect via the root tissue. Isolates of bean *Fusarium* from the UK have also been shown to be pathogenic on the roots of peas but the symptoms are not as severe as on *Phaseolus*.

Bean varieties differ in the degree of sensitivity to *Fusarium* foot rot but none are known to be completely resistant. Some progress with developing resistance or tolerance has recently become available through studies on root growth traits (Roman-Aviles *et al.*, 2004) and this together with soil management will be the most practical form of control.

Rhizoctonia root rot

This root rot is common throughout the world and is one of the most economically important root and hypocotyl diseases in beans. Losses occur in all types of cultivation, whether ploughed and cultivated, minimum tillage or direct drilling systems. Often the disease occurs in conjunction with *Fusarium* foot rot but the characteristic sunken reddish-edged lesions at the stem base are usually those of *Rhizoctonia solani*. It usually occurs in warmer weather and it is not as much a problem in northern Europe as *Fusarium solani*.

Resistance is available in breeding lines and some moderate resistance is present in commercial varieties; however, soil management is of prime importance to minimize the risks and to alleviate stresses on the root during the early part of the growing season.

Southern blight and white mould

Southern blight is serious in many tropical and subtropical areas of the world where high temperatures and wet soil occur through the growing season. The causal fungus is *Sclerotinia rolfsii* and it has a very wide host range in both monocotyledon and dicotyledons. including many vegetables. The early symptoms resemble those of its relative *S. sclerotiorum* except that most of the infection continues just above and just below ground, where the stem base is girdled and the plant dies. Sclerotia produced within the mycelium then serve as a source of inoculum for other hosts.

Deep cultivations where the soil is inverted bury the sclerotia and limit the contact between the sclerotia and the bean stems. Some resistant varieties are available and some progress has been made in the use of biocontrol agents such as *Trichoderma* and *Gliocladium* for use as antagonists to the fungus.

White mould caused by *S. sclerotiorum* occurs in most parts of the world, except the tropics, and infects a wide range of host crops. It has already been described as a disease in peas and much of the management that is used for the pea crop is relevant for beans (Fig. 6.12). Infection is triggered by periods of leaf wetness and optimal temperature; and ascospores are released from apothecia which in turn are produced from germinating sclerotia that have remained in the soil from the previous infected crop (Boland and Hall, 1987; Phillips, 1994).

Infection results in a collapse of the stem tissue and production of dense white mycelium, which covers the stems and leaves and infects the pods. Large sclerotia are produced within the infected tissue. The rate of infection is rapid and large-scale losses of yield and bean quality occur. There can also be

Fig. 6.12. *Sclerotinia* on green bean.

cross-infection from harvested fresh bean pods to healthy bean pods in stores or packaging, further reducing the saleability.

Some tolerance is known in breeding lines but few immune varieties are known and it is difficult to screen varieties for resistance as plant architecture can also influence infection (Schwartz and Singh, 2013). Biocontrol agents such as *Coniothyrium minitans* have been used and are successful in reducing the number of sclerotia. The inoculant needs repeated applications to build up populations and often the time between bean cropping in a field is too long for the agent to survive.

Botrytis pod rot (grey mould)

Botrytis cinerea is ubiquitous and present worldwide, infecting many crops and weeds during periods of high humidity. Infection occurs on the pod usually after first infecting the moribund petals that either adhere to the pod after pod set or fall on to the leaf petiole, where the fungus can then infect the stem, causing a stem girdling. *Botrytis* is also a very important post-harvest disease problem in fresh beans. The damage caused to the pods results in crop rejection of green beans for fresh market or for processing. In dry beans, the pods decay and the seed can be exposed, leading to blemishing, or where the seed is also infected the beans crumble.

B. cinerea is favoured by moist conditions and is more likely to be a problem if the weather is wet or if irrigation is applied to excess during the flowering and pod set stages, allowing the petals to remain adhering to the ends of the pods. A programme of fungicides can be used but in many cases the fungus has developed resistance to commonly used active ingredients.

Anthracnose

Anthracnose caused by *Colletotrichum lindemuthianum* is distributed worldwide but is more frequent in temperate and subtropical areas than in the tropics. It occurs in North and South America, as well as Europe, Africa, Australia and Asia. It is one of the most important diseases as it causes significant yield loss and damage through pod spoilage. It is seed borne and, once established, spores are spread by rain splash to surrounding plants very rapidly in warm wet conditions.

The symptoms of infection are very characteristic and black leaf lesions running along the leaf veins coupled with a shot-holing effect are seen at first. Later the pods become infected and deep circular lesions develop that are black in the centre but containing masses of pink pycnidia. Spores are then transmitted in rain splash to surrounding plants (Fig. 6.13).

Certain varieties are resistant to most races but the fungus is very variable in its pathogenicity. Seed-coat infections may be controlled by fungicide seed treatments but although foliar fungicides can be applied, the most effective means of control is by prevention through the use of healthy seed that has been produced in arid conditions.

Fig. 6.13. Anthracnose.

Bean rust

Uromyces appendiculatus occurs worldwide but is most common in tropical and subtropical areas. In Latin America it is most serious in Brazil, the Caribbean, Central America and Mexico and in Africa. It is rare in arid climates except under irrigation. Rust affects yield and pod quality for both the fresh vegetable market and for dry bean production and can reduce yield of dried beans by up to 30% through defoliation or reduction of leaf area. When pods become infected, the disease can cause deep dark pitting on the surface, making them unsaleable for the fresh or processing markets.

Symptoms first develop as small circular rust-coloured spots surrounded by a yellow halo. The pustules develop and produce orange-brown spores, which are transmitted to surrounding leaves and plants by rain splash or over-head sprinklers. Later, spots become larger and spores turn black.

A number of fungicides are available and are more effective if applied at the onset of the disease. The most effective management of bean rust is varietal resistance. Stable resistance is difficult, as there is a great degree of pathogenic variability in the fungus. Although most varieties are resistant to several races of rust, only a few have resistance to most local races. Many rust-resistant genes are present in the common bean and several have been used in developing rust-resistant germplasm and varieties (Park *et al.*, 2004). Substantial progress has been made in the USA in developing beans with multiple resistance genes.

Bacterial diseases

In all cases of bacterial diseases of beans, the primary source of infection is the seed and production areas for bean seed should be subject to strict phytosanitary procedures. Seed crops should be inspected during production in arid areas and tested for freedom of bacterial pathogens. In some situations a seed treatment containing streptomycin is used to provide surface-sterilizing properties but this is not available in all countries.

HALO BLIGHT   In temperate climates, halo blight caused by the bacterium *Pseudomonas syringae* pv. *phaseolicola* is the most common of the seed-borne bacterial diseases of beans. Leaf symptoms consist of small angular necrotic greasy spots on the leaf which become surrounded by a chlorotic area of tissue (Fig. 6.14). Later, sunken water-soaked grease spots appear on the pods and deepen into the pod wall until the developing seeds are infected. As the disease is easily spread during periods of rain or by overhead irrigation, the effect on the crop is usually first observed in small discrete patches that spread rapidly in the direction of the prevailing wind. Yield loss is caused by the defoliation and death of the plants but pod blemishing can result in rejection of the crop when harvested as fresh beans.

The source of infection is infected seed or crop debris but in a normal rotation the bacterium will not survive in the soil for longer than 1 year. There is no effective means of chemical control and therefore seed health is the only means of preventing the disease.

Fig. 6.14. Halo blight.

COMMON BLIGHT   Common bacterial blight is caused by *Xanthomonas campestris* pv. *phaseoli*. It affects the foliage and pods and is considered to be a major problem in most tropical or semi-tropical areas of bean production. The leaf symptoms appear as water-soaked spots that expand and the surrounding tissue becomes necrotic and bordered by a lemon-coloured area of tissue. Eventually the leaves take on a scalded appearance. Pod lesions are sunken and brown in colour. It can be very destructive in periods of warm humid weather, when losses occur in yield and bean quality.

BACTERIAL WILT   Bacterial wilt caused by *Curtobacterium flaccumfasciens* produces a wilting of bean plants during periods of moisture stress. It has been reported in the USA, Tunisia, Australia, Greece, Canada and Colombia as well as other production areas.

Viruses
A number of virus disease associated with *Phaseolus* beans have been described by Schwartz *et al.* (2005) but the most common serious viruses include bean common mosaic virus (BCMV), bean curly top virus (BCTV), bean yellow mosaic virus (BYMV) and cucumber mosaic virus (CMV).

BEAN COMMON MOSAIC VIRUS   BCMV can be found all over the world and in areas where susceptible varieties are grown the disease can cause serious yield loss, poor pod set and pod development. The virus can survive in weed hosts and to a small extent in infected seed. The transmission is then by aphid vectors, including *Acyrthosiphon pisum*, *Macrosiphon euphorbiaea*, *Myzus persicae* and *Aphis fabae*. The trifoliate leaves show irregularly shaped pale and dark areas over the surface and more commonly the leaves curl downwards and may grow longer than others. Some beans may develop a brown discoloration on the leaf veins and stems. Most commercial varieties are resistant to BCMV but dried bean types may be more susceptible. Efficient aphid control and planting before aphid activities begin will also reduce the risk of infection.

BEAN CURLY TOP VIRUS   BCTV occurs in western USA and in British Columbia. Seedlings develop as dwarfed plants, with severe puckering and downward rolling of the leaves. Plants are severely dwarfed and bunched. Leaves are brittle and flowers may abort. The virus has several hosts that are either perennial or winter annuals, such as Russian thistle or tumbleweed (*Salsola tragu*), mustard and also sugar beet. It is transmitted by the beet leaf hopper (*Circulifer tenullus*) and warm dry weather conditions early in the season favour migration from the overwintering hosts of the leaf hopper to newly emerging beans. There are a number of resistant varieties, as treatment of the leaf hopper is usually too late to be effective.

BEAN YELLOW MOSAIC VIRUS   BYMV develops in beans at any time before flowering. Leaves may be crinkled and pointed with some vein clearing. Leaves may droop, yellow mottling can occur on older leaves and plants may be stunted. BYMV is commonly distributed around the world and has several leguminous and non-leguminous hosts. The virus is aphid transmitted and infection may be on individuals or groups of plants. Depending on severity, the yield can be reduced and pods blemished or undersized. BYMV is also known as *Phaseolus virus 2* and can be seed borne in some legumes, including *V. faba*. Several aphid vectors can transmit the virus between beans from overwintering hosts. The principal vector is *Acyrthosiphon pisum* and the virus is persistent, having a wide host range including peas, clover, lucerne, gladiolus, freesia, lupin, soybean and wild legumes. There are some resistant varieties of dry beans available but early control of aphids in flowering crops is the most effective means of preventing wide-scale infection.

CUCUMBER MOSAIC VIRUS   Cucumber mosaic virus occurs in many countries, including in Europe and the Far East. In the USA, entire bean fields have been known to be affected but economic losses vary depending on the time of infection. When late infection occurs, pod quality is reduced.

Symptoms include narrowing and pointing of the leaves with a mosaic developing later on. Other symptoms include leaf curling, green and chlorotic mottling and dark green vein banding. Some strains of CMV can be seed borne in *Phaseolus* but the disease is readily transmitted by several aphid vectors. The virus has a wide range of weed hosts, including several perennial species.

Healthy seed is a useful means of prevention but management of aphids is more important. Some commercial varieties are tolerant to CMV and progress is being made to breed further for resistance.

SUMMARY

Infections of disease or infestations of pests are able to reduce yield or to adversely affect the quality of peas and beans for all the main markets. Many problems can be avoided by using a suitable crop rotation but although chemical control can be useful in many cases, effective application depends on the ability to define the optimum timing for applications of pesticides in order for them to work efficiently. With several pests and diseases, monitoring or forecasting systems have been developed and can be deployed in fields to give clear indications of the need to treat and the timing of such treatment. There is a growing requirement for the integration of management techniques to justify to customers the use of pesticides, and in many countries such practices are monitored by the end market.

For several diseases that are seed borne in origin, the use of clean tested healthy seed is an important start for production and current development in breeding for resistance into some varieties of peas and beans is helping to reduce the risk of crop loss and disruption to factory throughput.

REFERENCES

Adhikari, K., van Leur, J., Sadaque, A. and Trethowan, R. (2012) Breeding for rust (*Uromyces viciae-fabae*) resistance in faba bean (*Vicia faba*) in Australia. *Proceedings of the Sixth International Conference on Grain Legume Genetics and Genomics*, Hyderabad, India. CGIAR/ICRISAT, Patancheru, India.

Ali, S.M., Sharma, B. and Ambrose, M.J. (1994) Current status and future strategy in breeding pea to improve resistance to biotic and abiotic stresses. In: Muehlbauer, F.J. and Kaiser, W.J. (eds) *Expanding the Production and Use of Cool Season Food Legumes*. Kluwer Academic Publishers, Amsterdam, the Netherlands, pp. 540–558.

Allen, D.J. and Lennie, J.M. (1985) *The Pathology of Food and Pasture Legumes*. CAB International, Wallingford, UK.

Augustin, B. and Sikora, R.A. (1989) Methods for detection of *Ditylenchus dipsaci* infections in the seeds of grain legumes. *Gesunde Pflanzen* 41, 189–192.

Biddle, A.J. (1983) The foot rot complex and its effect on vining pea yield. *Proceedings of 10th International Congress of Plant Protection*, 117. British Crop Protection Council, Farnham, UK.

Biddle, A.J. and Cattlin, N.D. (2007) *Pests, Diseases and Disorders of Peas and Beans*. Manson Publishing, London.

Biddle, A.J., Blood-Smyth, J., Cochrane, J., Emmett, B., Garthwaite, D.G. *et al.* (1983) Pheromone monitoring of pea moth (*Cydia nigricana*). *Proceedings of the 10th International Congress of Plant Protection*, 161. British Crop Protection Council, Farnham, UK.

Biddle, A.J., Knott, C.M. and McKeown, B.J. (1988) *PGRO Pea Growing Handbook*. Processors and Growers Research Organisation, Peterborough, UK.

Biddle, A.J., Blood-Smyth, J.E. and Talbot, G. (1994) Determination of pea aphid thresholds in vining peas. *Proceedings of the Brighton Crop Protection Conference*, November 1994 2, 713–716. British Crop Protection Enterprises, Brighton, UK.

Biddle, A.J., Smart, L.E., Blight, M.M. and Lane, A. (1996) A monitoring system for the pea and bean weevil (Sitona lineatus). *Proceedings of the 1996 Brighton Crop Protection Conference – Pests and Diseases* 1, 173–178. British Crop Protection Enterprises, Brighton, UK.

Biddle, A.J., Thomas, J.E., Kenyon, D.M., Hardwick, N.V. and Taylor, M.C. (2003) The effect of downy mildew (*Peronospora viciae*) on the yield of spring sown field beans (*Vicia faba*) and its control. *Proceedings of BCPC International Congress – Crop Science and Technology*, 947-952. British Crop Protection Council, Alton, UK.

Boland, G.J. and Hall, R. (1987) Epidemiology of white mold of white bean in Ontario. *Canadian Journal of Plant Pathology* 9, 218–224.

Bond, D.A. and Pope, M. (1980) *Ascochyta fabae* on winter beans (*Vicia faba*): pathogen spread and variation in host resistance. *Plant Pathology* 29(2), 59–65.

Boulton, R.E. (1996) Pea early browning tobravirus. *Plant Pathology* 45, 13–28.

Bruce, T. (2016) Protecting pulses from pests – novel approaches to control. *Pulse Production and Protection*. A meeting held at Rothamsted Research, UK, March 2016. Rothamsted Research, Harpenden, and PGRO, Peterborough, UK.

Chen, Q. and Cai, L. (2015) *Didymella pinodella* (L.K.Jones). *Studies in Mycology* 82, 178.

Clarkson, J.P., Fawcett, L., Anthony, S.G. and Young, C. (2014) A model for *Sclerotinia sclerotiorum* infection and disease development in lettuce, based on the effects of temperature, relative humidity and ascospore density. *PLoS One* 9(4), e94049.

Fondevilla, S. and Rubiales, D. (2012) Powdery mildew control in pea. A review. *Agronomy for Sustainable Development* 32(2), 401–409.

Gladders, P., Ellerton, D.R. and Bowerman, P. (1991) Optimising the control of chocolate spot. In: *Aspects of Applied Biology 27, Production and Protection of Legumes*, pp. 111–116. Association of Applied Biologists, Warwick, UK.

HDC (2013) Reducing the impact of Sclerotinia disease on arable rotations, vegetable crops and land use. Project number FV361. Agriculture and Horticulture Development Board (AHDB), Kenilworth, UK.

Hillbur, Y., Bengtsson, M., Lofqvist, J., Biddle, A., Pillon, O. *et al.* (2001) A chiral sex pheromone system in the pea midge, *Contarinia pisi*. *Journal of Chemical Ecology* 27(7), 1391–1407.

Jellis, G.J. and Punithalingam, E. (1991) Discovery of *Didymella fabae* sp. nov., the teleomorph of *Ascochyta fabae*, on faba bean straw. *Plant Pathology* 40(1), 150–157.

Jönsson, B.G. (1988) An ecological approach to management of the pea midge, *Contarinia pisi* (Winn.), in vining peas, Department of Plant and Forest Protection, Swedish University of Agricultural Sciences. *Plant Protection Reports* 17. ISSN 0348-3428.

Kraft, J.M. (1994) Fusarium wilt of peas – a review. *Agronomie* 14(9), 561–567.

Kraft, J.M. and Pfleger, F.L. (2001) *Compendium of Pea Diseases and Pests*. American Phytopathological Society, St Paul, Minnesota.

Kumari, S.G. and van Leur, J.A.G. (2011) Viral diseases infecting faba bean (*Vicia faba* L.). *Grain legumes* 56, 24–26.

Linford, M.B. (1928) A Fusarium wilt of peas in Wisconsin. *Wisconsin Agricultural Experimental Station Bulletin* 85, p. 43.

Maurer, P., Couteaudier, Y., Girard, P.A., Bridge, P.D. and Riba, G., (1997) Genetic diversity of *Beauveria bassiana* and relatedness to host insect range. *Mycological Research* 101(2), 159–164.

Morgan, D., Walters, K.A. and Aegerter, J.N. (2001) Effect of temperature and cultivar on pea aphid (*Acyrthosiphon pisum*) life history. *Bulletin of Entomology Research* 91(1), 47–52.

Muehlbauer, F. and Chen, W. (2007) Resistance to Ascochyta blights of cool season food legumes. *European Journal of Plant Pathology* 119, 135–141.

Park, S.O., Coyne, D.P., Steadman, J.R., Crosby, K.M. and Brick, M.A. (2004) RAPD and SCAR markers linked to the Ur-6 Andean gene controlling specific rust resistance in common beans. *Crop Science* 44, 1799–1807.

Parke, J.L. and Rand, R.E. (1989) Incorporation of crucifer green manures to reduce Aphanomyces root rot of snap beans. *Annual Report of the Bean Improvement Cooperative* 32, 105–106.

Pfender, W.F. and Hagedorn, D.J. (1982) *Aphanomyces euteiches* f.sp. *phaseoli* a causal agent of bean root and hypocotyl rot. *Phytopathology* 72, 306–310.

PGRO (1970) *Annual Report of the Pea Growing Research Organisation*. Processors and Growers Research Organisation, Peterborough, UK.

PGRO (2015) *Recommended List of Bean Varieties. PGRO Pulse Growing Guide*. Processors and Growers Research Organisation, Peterborough, UK.

Phillips, A.J.L. (1994) Influence of fluctuating temperatures and interrupted periods of plant surface wetness on infection of bean leaves by ascospores of *Sclerotinia sclerotiorum*. *Annals of Applied Biology* 124, 413–427.

Pintureau, B., Gerding, M. and Cisternas, E. (1999) Description of three new species of *Trichogrammatidae* (Hymenoptera) from Chile. *Canadian Entomologist* 131, 53–63.

Roberts, I.M., Wang, D., Thomas, C.L. and Maule, A.J. (2003) Pea seed-borne mosaic virus seed transmission exploits novel symplastic pathways to infect the pea embryo and is, in part, dependent upon chance. *Protoplasma* 222(1–2), 31–43.

Rojas-Molina, M.M., Emeran, A.A., Fernández Aparicio, M. and Rubiales, D. (2010) Faba bean breeding for disease resistance. *Field Crop Research* 115, 297–307.

Roman-Aviles, B., Snapp, S.S. and Kelly, J.D. (2004) Assessing root traits associated with root rot resistance in common bean. *Field Crops Research* 86, 147–156.

Sanchez-Arroyo, H. (2015) *Featured Creatures*. University of Florida. Available at: http://entnemdept.ufl.edu/creatures/veg/bean/mexican_bean_beetle.htm (accessed 13 February 2017).

Schwartz, H.F. and Singh, S.P. (2013) Breeding common bean for resistance to white mold: a review. *Crop Science* 53, September–October, 1–13.

Schwartz, H.F., Steadman, J.R., Hall, R. and Forster, R. (2005) *Compendium of Bean Diseases*, Second Edition. American Phytopathological Society, St Paul, Minnesota.

Stawniak, N. (2011) Studies on stem nematode species (*Ditylenchus* spp.) associated with faba bean (*Vicia faba*) in the United Kingdom and their implications for field management. PhD thesis, University of Reading, UK.

Steenberg, T. and Ravn, H.P. (1996) Effect of *Beauveria bassiana* against overwintering pea leaf weevil, *Sitona lineatus. IOBC/WPRS Bulletin*19(9), 183–185.

Taylor, P.N., Lewis, B.G. and Matthews, P. (1989) Pathotypes of *Peronospora viciae* in Britain. *Journal of Phytopathology* 127, 100–106.

Tivoli, B., Avila, C.M., Banniza, S., Barbetti, M., Chen, W. *et al.* (2006) Screening techniques and sources of resistance to foliar diseases caused by major necrotrophic fungi in grain legumes. *Euphytica, Special Issue 147: Resistance to biotic stresses in legumes*, pp. 223–225.

Tu, J.C. (1991) Management of root diseases of peas, beans and tomatoes. *Canadian Journal of Plant Pathology* 14, 92–149.

Wade, B.L. (1929) The inheritance of Fusarium wilt resistance in canning peas. *Wisconsin Agricultural Experimental Station Bulletin* 97, 32.

Ward, R.L. and Smart, L. (2011) The effect of temperature on the effectiveness of spray applications to control bean seed beetle (*Bruchus rufimanus*) in field beans (*Vicia faba*). In: *Aspects of Applied Biology 106, Crop Protection in Southern Britain*, pp. 247–254. Association of Applied Biologists, Warwick, UK.

7

HARVESTING, NUTRITIONAL VALUE AND USES

As described in Chapter 1, peas and beans are used in a wide variety of ways, either fresh, where the immature pods or seeds are harvested and used as a vegetable, or processed, either by freezing or by canning, and as dried pulses as food ingredients or flour, or rehydrated and cooked. Cooked pulses may also be canned on their own or in mixtures, or processed in some other form. Dried pulses are also used in animal feed manufacture, either milled or heat processed with or without the seed coat, and fed to most types of livestock and in aquaculture. In all instances, the requirements for high-quality produce is important from a human health aspect but also economically in the processing: produce that is of poor quality is either unusable or will require cleaning and this will invoke payment penalties to the producer.

Each crop has its own particular set of operations to ensure an acceptable product, whether the product is consumed or marketed in a fresh state or processed at home or at a factory. In the case of dried pulses, the product must be harvested and stored in a safe environment before marketing.

In this chapter, each crop is considered firstly in the fresh vegetable form and secondly in the dried pulse form.

VINING PEAS FOR FREEZING OR CANNING

Maturity assessment

To ascertain the optimum time of harvest, there is a set of parameters that need to be fulfilled before the crop is taken. Pea maturity is measured by tenderometer prior to harvest. The produce of the vining pea crop is at its most acceptable stage for processing before the maximum yield has been achieved. During the early stages of maturation, when peas are at their optimum for freezing, yield is increasing rapidly; however, as the maturity advances to the stage where the peas may be suitable for canning, this increase begins to slow down as it approaches the maximum and then declines as moisture is lost. Beyond this the quality of the peas is judged to decline and therefore a sliding scale of payment to the grower usually operates. For this, the accepted measure of maturity is the tenderometer, though other assessment (physical, chemical or a combination of both) may be operated by the processor.

Maturation is accompanied by rapid changes in chemical composition that influence the flavour and texture of the peas. There is a loss of moisture and changes in the proportions of sugar, starch, cellulose, hemicellulose and pectin which, overall, increase the total solids. Sugar is converted to starch as maturity increases and so a measurement of alcohol-insoluble solids (AIS) can be used as an indication of maturity: the total percentage of AIS increases with advancing maturity and this has been correlated with assessments of quality that may be made by organoleptic methods. Such a laboratory process is time consuming.

The tenderometer provides a more practical means of maturity assessment, as it measures the physical aspects of pea quality in a repeatable manner. The measurement was derived from assessing the relative hardness of the pea by shearing peas in an enclosed cell and measuring the resistance of the shearing action. There is a close correlation between the physical and the chemical composition of vined peas and so the industry has adopted the tenderometer as a means of assessment which can be applied to vined peas at various stages of the harvesting sequence to allow an optimum time of harvest of the whole crop to be made.

Generally, the instrument has an upper grid of thin metal plates that can be driven by a motor through a second grid mounted on the same shaft. A sample of peas is placed in the space between the grids and when the instrument is set in motion the peas are first compressed, then squashed and finally forced through the lower grid. The force used is measured by the instrument and is displayed as a scale of 0–200 units, which is then adjusted to take temperature variations into account. The units are not easily defined, as they have been developed over many years of cross-checking between instruments and freshly vined peas of varying degrees of maturity.

In general, in the UK, peas are considered at an ideal maturity for freezing at around 100 units and those for canning at around 120 units. Because of the negative correlation of maturity with yield, usually the price paid to growers for higher-quality less mature peas is adjusted upwards.

Harvesting

The harvesting operation of vining peas is highly mechanized. Peas are usually sampled from fields that are approaching harvest and small samples are podded by hand or with a small vining machine in order to obtain a tenderometer reading. Readings may be taken daily 2–3 days prior to the crop being harvested in order to ascertain the optimum maturity. Complete pea harvesters are in operation throughout the world and combine the operations of picking, threshing and cleaning (Fig. 7.1). Pods are stripped from the stems and transferred by conveyors to the threshing drum, where peas are threshed from the pods; the peas fall through a set of screens that are kept clean by rotating brushes positioned along the length of the drum. The peas and some waste exit the machine through a further waste removal process and the waste is discharged on to the field. Peas are stored in a hopper prior to being unloaded into a trailer and taken from the field, usually directly to the factory.

Processing

Factories are usually situated within a 3 h delivery distance but where this is not possible the peas may be chilled on farm to 4°C before transportation, using iced water. Premium-quality peas are vined, delivered and frozen within 120–150 minutes.

On arrival at the factory, peas are sampled to ascertain the tenderometer value and the level of extraneous matter, which may include pods, stalks, insect or mollusc contaminants, stones, etc. Peas that fail to meet the quality standards may be refused by the factory and returned to the farm.

Fig. 7.1. Harvesting vining peas.

The peas are tipped into holding bins, in which they are moved by elevators to the rest of the cleaning processes. These involve compressed air blowing and washing in tanks designed to float away light trash and allow stones to sink for later removal. Clean peas are then either frozen or canned.

For freezing, there are two types of commercial processes: vibratory bed systems and a belt system. Vibratory bed freezers push refrigerated air through holes in the floor of the bed and the angle of the holes allows the peas to be moved in a wave motion to the freezer outlets. Some beds also vibrate to keep the peas individually quick-frozen. In the belt freezer, peas are frozen on a wire mesh belt through which cold air is forced as the belt is driven through the refrigerated freezer box. Frozen peas may be optically sorted at this stage or can be screened over vibrating separators to remove any defects (PGRO, 2015).

The canning operation takes place immediately after cleaning and blanching. Peas are scalded by steam to prevent deterioration by microorganisms and then put into cans where brine solution, which may contain sugar, salt and water, is added. The cans are sealed and cooked in a pressurized cooker for 20 min at a temperature of 121°C. The cans are then cooled before labelling and packing.

Frozen peas are usually grown on contract to the processor and they are supplied on an agreed tonnage basis on a daily schedule for the duration of the processing season, which may be up to 6 weeks. In the UK, a reliable supply to the factories is ensured from the producers, who are usually formed into grower cooperatives sharing drilling, agronomy and harvesting equipment. Each cooperative or grower group organizes seed supply, drilling schedules and agronomy and a shared set of harvesters and transport is able to deliver peas to the factory on a regular basis on a 24 h, 7-days-a-week schedule. Peas are generally marketed by the processor to retailers and food service companies, who in turn supply the catering and allied trades. Payments to grower cooperatives or groups are based on a shared risk in that growers receive a fixed payment per tonne of frozen product regardless of the variety or time of harvest. Payment is usually agreed between the processors and the producers at the beginning of each season. Demand for frozen product varies over the year and frozen peas may be held in freezer storage for many months. Occasionally there is a carry-over of stock when demand is low or the yields in a particular year are high and this affects the contracted area for the following year.

FRESH MARKET PEAS

Maturity assessment

Peas are harvested by hand to minimize pod damage. The pods are stripped from the stems in the field and packed in boxes before being transported to the farm or pack house for packing (Fig. 7.2). The optimum time for harvest may

Fig. 7.2.   Picking peas for fresh market.

be assessed by tenderometer but, in general, an assessment of the percentage of shelled peas to pod is used on a wider scale. When harvesting time has been reached, the pods are usually removed in a single destructive harvest by pulling out the plant at ground level and hanging it upside down to make the pods more easily visible.

Packing

Pods are then packed into field boxes and transported to the farm packing shed. Boxes are moved quickly into a chiller to remove field heat and to reduce dehydration. Usually the farm chilled stores are set at 8–10°C. This is particularly important when the peas are grown in warm areas. The boxes are transported to the main pack house, where the product is checked by quality control inspectors before it is moved for further chilling to 5–8°C to await final packing.

Evenness of product such as pod length, average number of seeds per pod and shape of pod are the main criteria but also the presence of a string along the pod suture is undesirable for the mangetout and snap pea markets. Pods should be clean and free from blemish or physical damage. The colour is also important for edible-pod types. The pods should be fully turgid and clean. The stem and calyxes should be green and there should be no blossoms attached to the pods, as this can lead to rot spots and fungal breakdown issues. The product is not stored in high humidity, as this would encourage fungal breakdown and rotting.

The pods are then packed into smaller trays for retail sales and covered with film to allow some air movement. Usually distribution to supermarkets takes place using chilled transport at 7–10°C. At the retailers, most are displayed in chiller cabinets that are set at around 15°C. Edible-pod peas are highly perishable and will not maintain good quality for more than 2 weeks. Wilting, yellowing of pods, loss of tenderness, development of starchiness and decay are likely to increase following storage beyond 7 days; longer-term storage of edible-pod peas is usually at 1–3°C to slow down respiration rates.

BROAD BEANS FOR FREEZING AND CANNING

Maturity assessment

Maturity of broad beans is assessed in a similar way to that of vining peas, using a tenderometer. The tenderometer is not, however, an absolute guide to quality, since the reading is in fact an average of all the beans in the sample, which may contain beans of widely different maturities. It has also been established, by organoleptic tests, that the optimum quality for a particular variety may occur at different maturities when grown under different conditions; similarly, optimum quality may occur at different maturities in different varieties. Under dry conditions the skin may toughen more rapidly as the chemical processes take place within the bean. In the absence of more critical methods, a combination of tenderometer values and organoleptic tests is often used.

Harvesting

Beans are usually harvested directly in the field by vining machines equipped with modified picking reels to deal with a taller crop, or they may be first cut and left in windrows for 24 h to allow the crop to wilt slightly, which in some situations can assist the threshing operation.

Beans are unloaded from the mobile viners into tanks for transport either direct to the factory or to an on-farm cleaning and cooling line. They are then transported in sealed bins to reduce discoloration of the vined beans. The beans are cleaned again in the factory before blanching and freezing. Beans may also be canned in the same way as peas. In extreme cases of shortage of suitable crop close to the factory, canning can take place using frozen broad beans. As with frozen peas, beans are produced on contract to the processing companies and sales are usually made by the processor to the retailers.

Harvesting broad beans for fresh market

Broad beans for hand picking are usually grown in wider rows to enable access for the pickers. The assessment of maturity is sometimes made by tenderometer but often the criteria for harvest are based on visual inspections of the pods, their length and degree of pod fill, and samples are taken to ascertain the seed size and the seed-to-pod ratio in terms of weight. The fresh market requires a pod width and length of defined dimensions, a minimum number of beans per pod and a seed-to-pod ratio of around 35%.

Bean pods are picked by hand and transported to the packing house in boxes before pre-packing and transport to retailers.

PHASEOLUS BEANS FOR FREEZING

Maturity assessment

Green beans are at their most acceptable stage for processing before maximum yield has been obtained. Yield increases rapidly during the early stages of maturation, when the crop may be harvested for freezing, while yield increases more slowly during the later stages, when the crop may be harvested for canning. While it is not usual to have a sliding scale of payment to the growers based on crop maturity, it is usual to have an acceptable means of assessing maturity to enable standards to be maintained. The main stages of bean maturation can be summarized as: (i) rapid increase in pod length with relatively slow seed development; (ii) enlargement of the pod and more rapid enlargement of the seed; and (iii) lignification, senescence and drying of the pod and drying and hardening of the seed.

Several methods have been developed that measure the physical changes taking place during maturation and generally these form a more practical means of determining maturity than any chemical assessments. A common method of assessment is based on average seed length. The seed length measurement is obtained by measuring the total length of ten seeds, each being the largest seeds taken from the largest pod taken from a ten-plant sample. Depending on the variety of bean used, a practical stage for freezing can be taken between 100 mm and 120 mm of seed length. For canning the seed length is extended.

Harvesting

Green beans or snap beans are grown on a large scale commercially and all the crop for processing is mechanically harvested. Beans are drilled in the spring, as a mean soil temperature of 10°C is required before germination can commence. Beans are usually grown in beds 2 m wide containing up to four rows of plants, though the harvesters are able to travel along or across rows if necessary. Harvesting is conducted by a self-propelled bean harvester equipped with longitudinal picking reels that revolve along the rows and the combing action of the spring tines attached to the reel removes the pods from the plant. These are elevated into the machine together with a certain amount of leaf and other debris. The pods are then elevated to the cleaning and declustering sections of the harvester. The light debris (primarily leaf and stem) is blown back on to the field. Devices for separating the clusters of pods are capable of dealing with a high percentage of these and the remains are also blown back on to the field. The beans are then offloaded into boxes and transported to the factory for processing.

The freezing and canning operations are similar to those described for peas and broad beans but preparation of the pods by either cutting, slicing or snibbing the pod ends to produce a 'whole pod' pack takes place prior to the steam blanching process.

Phaseolus beans for fresh market

Many of the operations used for producing fresh peas and broad beans are carried out to ensure the beans are of acceptable quality for the fresh market.

DRIED PEAS AND BEANS

Pulses are regarded as a beneficial source of nutrients and are recommended as a staple food by health organizations and dieticians. They are rich sources of vitamins, minerals and carbohydrates in the human diet and a useful source of protein and energy in livestock rations. They represent an important source of protein for vegetarians and have a low glycaemic index (Rizkalla *et al.*, 2002). Pulses are important foodstuffs in tropical and subtropical countries, where they are second in importance only to cereals as a source of protein. They are also recognized as a food choice with significant potential health benefits. They contain complex carbohydrates (dietary fibres, resistant starch and oligosaccharides), protein with a good amino acid profile (being high in lysine), important vitamins such as B vitamins, and folates and iron as well as antioxidants and polyphenols.

Pulses are harvested, handled and processed in a variety of ways and undergo several primary and secondary processes such as cleaning, de-hulling, puffing, grinding and splitting prior to their consumption. There is a very small increased interest in many developed countries in the utilization of whole pulses and their milled constituents, particularly with flours and fractions such as protein, starch and fibre, and using these in food products. There are new developments in the use of processing techniques such as extrusion cooking including these ingredients to produce speciality foods such as pasta products, baby food, snack foods, dried soups and pet foods. In addition, research is under way to exploit pulse ingredients in pharmaceutical or 'nutraceutical' products (Carbonaro, 2011).

Peas and beans for dry harvest require specific harvesting criteria and the crops are considered separately.

Combining peas

Harvesting
The crop is harvested when fully mature but in some crops, where there has been uneven maturation or areas of weeds that have delayed maturation, the haulm can be desiccated chemically to allow harvest to take place. Desiccation involves applying a chemical to prevent further growth of the crop and weeds and is followed by direct combining when the peas have dried sufficiently. Desiccation can cause the crop to lodge and there is a risk of collapse of the stem due to insufficient lignification if applied too early. Other effects include shrivelled seeds; and if left in the field beyond the optimum time for harvest, pod shatter can occur where the peas are discharged from the pods and fail to be collected by the harvester. The crop is generally harvested by the combine harvester without the need to cut the crop first. It is preferable to combine the crop when the peas pass through without damage, when the dry matter has reached 75% (Fig. 7.3).

Fig. 7.3.   Harvesting combining peas.

They may then be dried artificially. Damage to the testa can occur at very low moisture contents of the seed; and if left in the field for too long, pod shatter and seed loss can occur in the field. Peas for the high-quality human consumption market can bleach very easily, particularly if left in the field when rain is followed by sunshine.

Processing

Once harvested, peas may be kept on farm or delivered to grain elevators for long-term storage. Before use, extraneous matter must be removed. Many grain merchants or cooperative elevator companies have large-scale machinery for processing dried peas before shipment to food processors, packers or animal feed mills. The first process involves passing the peas over screens to take out soil, stones or larger extraneous matter. These are usually gravity separating screens, which are slightly angled, and the peas are vibrated over the screens so that undersized seeds and lightweight waste move to the upper edges of the screens and are removed. The next operation is to pass the peas through a bank of electronic colour sorters. Peas are dropped through a beam of light calibrated to activate a jet of compressed air when the reflectance of the pea falls out of specification, i.e. lighter or darker due to bleaching or stain. Once activated, the air jet blows the blemished pea sideways into an adjacent waste-collecting bin. Each bank of sorters can typically process around 12 t/h.

A further stage of cleaning may be necessary if the peas have been damaged by insects such as pea moth larvae (*Cydia nigricana*) or pea bruchid (*Bruchus pisorum*), both of which leave holes in the seed. This process involves passing the crop through a horizontal cylindrical drum lined with fine needles. As the peas pass round the drum, damaged peas are impaled on the needles and are then brushed off on to a collecting conveyor belt to the waste bin. The pin drum or needle drum may be operated in a bank of drums to speed throughput.

Finally the peas are packed into either bulk bags or smaller paper sacks before being delivered to the shipper or processor.

Peas for animal feed may be milled and pelleted, or green-seeded peas can be micronized to produce a flaked pea. Micronizing involves passing the peas under a high-temperature heater before rolling them flat and allowing them to cool. It is thought that this process makes the peas more easily digestible. The micronized product can also be used as an ingredient in pet food.

Uses

Marrowfat peas for human consumption can be sent for further processing. In the UK and some European countries, marrowfat peas are canned as 'processed peas'. This involves soaking the dried product in water for 18 h before draining and steam blanching. The blanched peas are put into cans with a brine solution before sealing and are cooked for 21 min at 121 °C before cooling, labelling and dispatch to a distribution warehouse.

Peas may also be sold whole and dry for home consumption in small packets. A larger market exists for de-hulled marrowfat peas, which are then soaked and boiled and eaten as 'mushy peas', a delicacy common in the UK and Australasia as an accompaniment with fish and chips.

A number of processes for peas also include their use as a snack food where the peas are deep fried and flavoured. Pea flour is used in the baking industry and the yellow-seeded varieties are preferred for this. Pea flour is useful in certain biscuit and batter recipes and, because it is gluten free, pea flour is suitable for coeliac diets.

Peas are either grown on contract to a merchant or they may be sold on the open market. In Europe, often the peas are grown by farmers who are members of a grain-producing cooperative and the peas are stored in bulk until delivery to the end user. Quality of peas is very important, especially for the human consumption market, and prices on the open market are not fixed in the same way as cereals that can be sold forward at a known price. Prices can be volatile, depending on demand. Production in other countries may compete for the same markets and prices reflect the supply and the quality. UK production of high-quality marrowfat peas can be affected by adverse weather at harvest and peas of high quality can then demand high premium prices. Overproduction of pulses in Canada, for example, can also deflate prices paid for animal feed.

Faba beans

Harvesting

Beans may be either combined direct or from a windrow; the majority of crops are combined directly and without the use of a chemical desiccant. Bean pods blacken and seeds become dry and hard first, the stems of most varieties remaining green for several days after this. In some seasons, pod ripening is uneven and upper pods remain green for longer. The pods will be easily threshed and the seed fit for combining when the seed moisture content is between 16% and 20% (Fig. 7.4). Crops that are too dry can shatter, resulting in high combine losses. In very dry seasons when beans are combined at very low moisture contents, seed-coat cracking or seed splitting may occur and quality of crops grown for seed may be reduced.

Once harvested, the beans may be stored satisfactorily at 14% moisture content. Where drying is required the method used is dependent on the market. Field beans grown for seed or premium niche markets require drying and storage so that the germination and quality are not affected by damage such as seed-coat cracking or seed splitting. For the animal feed market, maximum moisture content, admixture levels and freedom from moulds are the only criteria stipulated and the beans do not require such gentle drying. The meal may be pasty if insufficient moisture is removed from the centre of the bean, and if stored too long it may become rancid.

Fig. 7.4. Mature faba beans.

Processing

Bean cleaning before use is usually carried out by specialist merchants and the procedures are similar to those used in the cleaning of combining peas. Staining, broken and extraneous matter is removed. Where beans are destined for human consumption, the beans are cleaned in a pin drum cleaner to remove beans that have been damaged by the emerging bean seed beetle (*Bruchus rufimanus*).

Uses

There is a range of uses for faba beans. The principal use is as a source of protein and carbohydrate for livestock. Recent studies have shown that beans can be used to substitute soybean meal in diets for pigs as well as in beef production. Despite the presence of tannin in the seed coat of beans, experiments have shown that there are no detrimental effects on feeding a relatively high proportion of beans to pigs from the grower to the finisher stages of pork production (Houdijk *et al.*, 2013) but because the bean amino acid profile is short of the sulphur amino acids, some balance has to be made using synthetics.

There is an increasing demand for beans in Scandinavian and Scottish aquaculture. Beans are de-hulled and used to manufacture compound feed pellets for salmon farming. The feed has to be adjusted using synthetic amino acids together with additional fishmeal but the beans are well suited to hold the pellets together once immersed in water and the specific gravity allows the pellets to sink slowly, giving maximum opportunity for the fish to feed. The hulls are then pelleted and marketed as a source of fibre for the animal feed industry.

Beans are also a staple diet in the Middle East and East Africa. There is a significant proportion of beans grown in Egypt and Sudan but current consumption of 500,000 t of beans per annum cannot be met from local production. Large-seeded pale-skinned beans are required for this market and these should be blemish free and with a thin intact seed coat. High-quality beans from the UK, France and Australia are exported to the Middle East each year. The beans are soaked and boiled whole with spices as 'breakfast beans' (ful mesdames) or ground to make falafel. They are an ideal slow energy-release food source, especially during the period of Ramadan when food must only be consumed during the hours of darkness.

As with peas, the supply of beans is dependent on the demand and the price is dependent on the quantity and quality. The market for human consumption is affected by strong competition from the UK, France and Australia and the weather in the different areas of production can have a direct adverse or favourable influence on the amount of product available.

Dried beans

Harvesting
Beans are harvested when the crop is fully mature. The crop may be cut and left in a swath before pickup by the bean combine harvester or it may be cut directly. In most American and Canadian production the beans are undercut, which slices the root off and lays the plant on the ground. The beans are then windrowed into a swath, which shakes off some of the soil and helps to cushion the beans during harvest. When the beans dry down to 20% moisture content, the windrows are picked up by a combine harvester. The combine harvester is adjusted to cause minimum damage to the beans. Bean seed is extremely fragile and crops are harvested early in the day to reduce the amount of seed-coat or shelling-out losses. Beans can be damaged when handled below 13% moisture content. Often small-scale production is harvested by hand, either by pulling the haulm and spreading on the ground to dry completely or hanging vines in stacks for drying. Threshing and sorting is often carried out by hand-winnowing methods.

Uses
Because of the presence of trypsin inhibitors in the beans, they are not suitable for animal feed; and for human consumption it is necessary to cook them

before eating. They are a very high source of protein and are eaten widely in many developing and underdeveloped countries. Dry seed requires relatively simple soaking and cooking before eating. To reduce cooking time, beans are pre-soaked before boiling or pressure cooking and are consumed either as a soup, purée or plain without the broth. Like most pulses, they are low in the sulphur amino acids such as methionine and to provide an adequate diet they are consumed with other vegetables or cereals such as rice.

Dried beans have many uses but as a processed crop the white navy bean is probably the most well known as canned baked beans. Beans are soaked and then pressure cooked in the can with tomato sauce (a mixture of corn flour, spices and tomato purée). The UK is one of the largest importers of navy beans, with more than four million cans of baked beans being produced each day by two factories in England.

Attempts have been made to develop the navy bean as a crop in the UK. The main constraint in this development is the shortness of the growing season and the risk of unfavourable weather conditions adversely affecting the completion of maturation and subsequent harvesting. The suitability of growing areas is also restricted in that beans have a low tolerance to cool temperatures. Sowing is best carried out when the average soil temperature has reached 10°C and this may not occur until early to mid-May. Similarly, cool temperatures during the growing season shorten plant height and the resulting mature pods are too close to the soil level for successful conventional combine harvesting. Work progressed in developing earlier-maturing varieties that were more likely to be successful and which also had resistance to seed-borne bacterial diseases such as halo blight. However, the economics of production were such that the relatively low yield, the uncertainty of success and lack of investment in specialist harvesting equipment and the demand from the processors for large quantities of beans to be available to process all year round were not conducive to any significant level of UK production. In contrast, a few other types of *Phaseolus* beans, especially coloured-seeded types, have been developed and grown in the UK on a small scale (Leakey, 1999).

Other beans, especially the dark red kidney and large butter beans, are canned in brine. Other processes include rehydration and freezing whole beans to produce a bean mix salad. For domestic consumption beans are soaked and cooked either in mixed-bean stews or de-hulled and split for use as dhal.

NUTRITIONAL COMPOSITION OF VEGETABLE PEAS AND BEANS

All commodities are either sold as a single product, or peas and beans are commonly produced in mixes as frozen or canned products. Frozen peas are often mixed with frozen cooked rice or sweetcorn and broad beans may be mixed with peas or edamame soya or lima beans as a frozen or canned product.

Typical nutritional values derived from a range of types for frozen peas or broad beans are shown in Table 7.1.

NUTRITIONAL COMPOSITION OF DRY PULSES

For human consumption, the various processes for preparing and cooking may change the nutritional value of the dried pea or bean but generally it is accepted that the consumption of pulses as part of the diet is highly beneficial. As a guide to the values for human consumption, Table 7.2 shows comparisons between the main pulse crops described in this chapter.

As ingredients or components for animal or fish feedstuffs, peas and *Vicia* beans are well known sources of protein and energy, although low in some of the sulphur amino acids. Peas and faba beans have a total protein content between that of soybean meal and cereals. As for all raw materials, the total

Table 7.1. Typical nutritional value of frozen peas and green beans (data from British Growers Associataion and USDA Standard Reference SR27).

100 g serving (boiled, no added salt)	Frozen peas	Green beans	Broad beans
Calories	327 kJ	130 kJ	461 kJ
Protein	5.2 g	1.8 g	5.6 g
Total carbohydrate	14.3 g (4.7 g as sugar)	7.1 g (1.4 g as sugar)	11.7 g (1.8 g as sugar)
Dietary fibre	5.5 g	3.4 g	4.2 g
Sodium	72 mg	6 mg	50 mg
Vitamin C	9.9 mg	16.3 mg	33 mg

Table 7.2. Nutritional values of dried peas and beans for human consumption (data from USDA Standard Reference SR21).

100g serving	Dried peas (boiled)	Faba beans (boiled)	Dried *Phaseolus* beans (boiled)
Calories	494 kJ	461 kJ	494 kJ
Protein	8.3 g	7.6 g	8.3 g
Total carbohydrate	21.1 g (2.9 g as sugar)	19.7 g (1.8 g as sugar)	21.1 g (2 g as sugar)
Dietary fibre	8.3 g	5.4 g	7.0 g
Sodium	2.0 mg	5.0 g	2.0 g
Iron	1.3 mg	1.5 g	2.1 g
Folate	65 µg	107 µg	102 µg

crude protein (calculated as nitrogen $\times$ 6.25) is variable between crops, the growing environment and year; however, on average, the protein values remain standard and set values are used by feed manufacturers for peas at around 21% and for faba beans at around 25%. In surveys of commonly grown varieties of peas and beans (spring or winter) there has been very little difference in protein levels between varieties of peas (Bastianelli *et al.*, 1995) or peas and beans (Houdijk *et al.*, 2013).

An analysis of the composition of peas and beans compared with soybean meal is shown in Table 7.3.

The composition of the amino acids in peas and beans is dependent on the total protein content and the proportion of the different proteins. Lysine levels are intermediate between that of soybean meal and cereals which is nutritionally essential for non-ruminants, but peas and beans do not contain adequate levels of sulphur amino acids or tryptophan (Table 7.4).

Starch is the most abundant component of peas and beans at around 500 g/kg and is a valuable energy source for livestock. The mean oil content is low (less than 2%). The composition of this oil is similar to that of cereals, being mostly triglycerides, polyunsaturated in nature with a predominance of linoleic acid (see Table 7.2).

In Europe, most commonly grown white-flowered peas contain only low levels of trypsin inhibitors, though occasionally varieties have been found to contain levels of between 7 mg and 10 mg, which is not considered to be too high for animal feeds. In faba beans, most varieties contain tannins but, as

Table 7.3. Nutritional composition of peas and beans for animal feed.

	Combining pea	Faba bean	Soybean (Hy-Pro)
Dry matter %	86	86	86
Oil %	1.0	1.5	1.9
Starch %	44.6	37.0	–
Cellulose %	5.2	8.0	6.0
Minerals %	3.0	3.5	6.4
Protein %	20.7	25.0	54.3

Table 7.4. Protein and amino acid composition (Houdijk *et al.*, 2013).

	Crude protein	Lysine	Methionine	Methionine + cystine	Threonine	Thiamine	Valine
Combining pea	20.7	1.51	0.91	0.48	0.79	0.20	0.97
Faba bean	24.7	1.56	0.18	0.48	0.85	0.21	1.11
Soya bean	47.78	2.92	0.65	1.35	1.86	0.65	2.27

stated earlier, there is little evidence to show that this is high enough to affect growth in pigs (Houdijk *et al.*, 2013)

Many studies have examined the value of peas and beans as a protein source for livestock and it is accepted that peas and beans may be used in nutritionally balanced diets for many ruminants and non-ruminants and may completely replace soybean meal without detrimental effect on livestock performance. In addition, the increasing use of de-hulled faba beans in aquaculture for salmon in Norway and Scotland will gradually replace the need for fishmeal in the rations.

The value of straw

Straw disposal after combining peas can be a problem in some instances but there is an increasing use of pea straw for use as feed for both sheep and cattle. With a higher protein content and less fibre, pea straw has a higher nutritive value than that of cereal straws. Its quality is intermediate between a cereal straw and a good grass hay (Mould *et al.*, 2001; Ellwood, 2004). The nutrient value of pea haulm is shown in Table 7.5.

SUMMARY

Whether for human consumption or animal or fish feedstuffs, peas and beans are a useful source of protein and energy, whilst as vegetables they provide a source of vitamins. There is a developing market for pulses in animal and aquaculture production and this demand is increasing as the supply of animal- or fish-based feedstuffs declines and the growth in consumption patterns for meat and dairy products, particularly in developing countries, increases demand and use of soybean for feed. However, peas and faba beans do not contain a comparable level of sulphur amino acids as does soybean meal, and although protein and energy values are acceptable for animal production,

Table 7.5. Nutrient value of pea haulm (%) for animal feed.

Component	%
Dry matter	88.8
Crude protein	8.2
Crude fibre	36.3
Lignin	7.2
Ash	9.8
Gross energy	18 MJ/kg

especially for pigs, diets and formulations containing peas or faba beans require the addition of synthetic amino acids. As with all crops, attention to the harvest operation and the correct time of harvest is essential to maintain the quality required for the various markets. Some of the methodology in assessing maturity is based on long experience whilst developments in processing of both the raw vegetable and the dried products ensure high-quality clean acceptable produce for consumption. There are few other commodities that are consumed in the natural state in that both seed coat and seed itself are eaten, usually after cooking. It is essential, therefore, that adequate care is taken to maintain the quality and demand for these crops.

REFERENCES

Bastianelli, D., Carrouée, B., Grosjean, F., Peyronnet, C., Revol, N. and Weiss, P. (1995) *Peas – Utilisation in Animal Feeding*. UNIP-ITCF, Paris.

Carbonaro, M. (2011) Role of pulses in nutraceuticals. In: Tiwari, B.K., Gowern, A. and McKenna, B. (eds) *Pulse Foods*. Academic Press, London, pp. 385–418.

Ellwood, L.S. (2004) *Research Summaries: Peas in Livestock Diets*. Pulse–Canola Feed Literature Database, Saskatchewan Pulse Growers, Saskatchewan Canola Development and InfoHarvest Inc., Saskatoon, Canada.

Houdijk, J., Smith, L., Tarsitano, D., Tolkamp, B., Topp, C. *et al.* (2013) Peas and faba beans as home grown alternatives for soya bean meal in grower and finisher pig diets. In: Garnsworthy, P.C. and Wiseman, J. (eds) *Recent Advances in Animal Nutrition*. Nottingham University Press, Nottingham, UK, pp.145–175.

Leakey, C.L.A. (1999) Progress in developing dry *Phaseolus* beans for Britain. In: *Aspects of Applied Biology* 56, *Protection and production of combinable crops*, pp. 195–202. Association of Applied Biologists, Warwick, UK.

Mould, F.L., Hervas, G., Owen, E., Wheeler, T.R., Smith, N.O. and Summerfield, R.J. (2001) The effect of cultivar on the rate and extent of combining pea straw degradability examined *in vitro* using the Reading Pressure Technique. *Grass and Forage Science* 56(4), 374–382.

PGRO (2015) *PGRO Vining Pea Growing Guide*. Processors and Growers Research Organisation, Peterborough, UK.

Rizkalla, S.W, Bellisle, F. and Slama, G. (2002) Health benefits of low glycaemic index foods, such as pulses, in diabetic patients and healthy individuals. *British Journal of Nutrition* 88, 255–262.

THE FUTURE FOR PEAS AND BEANS

INTRODUCTION

In the earlier chapters, aspects of current development in pea and bean breeding and production have been discussed, albeit mainly concentrating on large-scale developed commercial agriculture in the developed countries of the world. However, on a general note, it has been established that of all the large-seeded legume crops, peas and beans are the most versatile and are able to grow in a wide variety of geographical areas and soil types. Some species are more adaptable to conditions than others but within species there are many commercial varieties and types that perform well in most situations. These facts alone make the future of these crops much more certain.

As a food source, peas and beans are well accepted for animal nutrition and for human food with relatively little processing of the raw ingredient necessary in a wide variety of cases. In human nutrition, pulses have been shown to have an important role in preventing illnesses such as cancer, heart disease and diabetes (Pulse Canada, 2008). The dry edible seeds of large-seeded legumes, known as pulses, contain a higher level of protein than cereals and high levels of both soluble and non-soluble fibre with a low glycaemic index. In developing countries, pulses are an important source of vegetable proteins and constitute the main source of protein for most populations.

As a crop, the role of legumes in the farm rotation has become even more important as the pressure to reduce the use of chemical fertilizers and pesticides becomes greater in many areas of the developed and developing world. Their value as a non-cereal crop grown in a rotation to reduce the risk of soil-borne diseases and weeds and the non-reliance on the use of chemical nitrogen fertilizer is a major advantage. It has now been established that growing legumes has an overall effect in reducing greenhouse gas emissions over the whole farming system, as the production of nitrous oxide is much lower in the absence of the requirement for nitrogen fertilizer (Jeuffroy et al., 2013).

The demand for meat and dairy-based products in developing countries has increased significantly over recent years and this is reflected in the demand for soya protein for animal feed. For example, there has been a rapid increase in the imports of soya by China from around 40 million tonnes in 2006 to around 80 million tonnes in 2016 (source: US Department of Agriculture), which has the effect of decreasing the availability for the rest of the world and increasing the price of the product. There has been a net increase in the use of pulses for feed in Europe and other areas that are unsuitable for soya production and this, coupled with national incentives available to farmers, has led to an increase in production of peas and beans.

Demand and supply is therefore increasing mainly due to these factors but future development of the crops is likely to be affected by other additional factors. Some of these involve constraints in production due to a range of agronomic situations, including climate, available land and agrochemical inputs. Limitations on crop performance due to yield plateauing and susceptibility to pests and diseases also have an important role affecting the development of legume crops. Other factors include: an increasing use of the crops for food and animal feedstuffs; encouragement of consumers to replace meat-based food with vegetable protein and increased awareness of its health benefits; developments in processing of the crops; increasing the diversity of products and convenience foods; and novel food products. The development of technology to extract components of the harvested crop, particularly dried crops, and utilize these in food, pharmaceuticals and industrial products will all have an impact.

This chapter will further examine some of these factors and discuss them in relation to the future of peas and beans.

CONSTRAINTS AND OPPORTUNITIES

Varietal performance

The continued development of new varieties by breeders is dominated by the potential of high yield, produce quality and resistance to pests and diseases. Inherent in these major attributes are other factors including harvestability, e.g. standing ability of the crop and resistance to adverse conditions such as drought or temperature changes.

Legumes are naturally sprawling or climbing plants, characteristics that are not suited to large-scale mechanical harvesting. Breeders have selected types that are generally short in stature and have relatively good standing ability, at least during clement weather conditions. Straw stiffness is not a natural characteristic, unlike the cereal crop, and therefore efforts to select for this attribute have concentrated on stem thickness and, in peas, the ability to self-support through the reduction of leaflets and their replacement with tendrils.

With faba beans the development of shorter stems has much improved the standing ability of the crop, particularly as it matures and reaches the harvest stage. Ease of harvest by machines has been improved by design features such as modified crop lifters that are positioned at the leading edge of the cutter bar at soil level: as the machine travels forwards, crop that has lodged is raised upwards over the cutters and on to the pickup reel. Lifters especially developed for lodged pea crops have been in use for some time but improvements continue to be made by the manufacturers.

Yield

Most modern pea varieties have the potential to set two or three pods at each reproductive node, though two pods are more commonly found. The number of reproductive nodes is usually defined by the variety. The total number of pod-bearing nodes is therefore set.

Factors affecting pod set have been discussed earlier and these include nutrition or drought stress. In an optimum growing situation the potential for maximum yield can be achieved and the development of varieties with multi-pod character has the additional potential for yield increase.

With faba beans the propensity to abort flowers during the pod-setting stage is greater than in peas and current work is examining the mechanisms involved in pod set as a means of increasing yield.

Recent studies have indicated that the introduction of new varieties of faba beans and their commercial uptake has, in the UK, resulted in a steady increase of yield, typically from around 3.7 t/ha in 1963 to 5 t/ha in 2014 (NIAB TAG, 2014).

Quality

Breeders of peas and beans have quality as one of the most important attributes in a new variety. In the vegetable sector, quality infers palatability, colour, taste and size or shape of the harvested produce. In peas, those characteristics for greenness and sweetness have been identified and breeders' material is being used to develop new varieties (Domoney, 2011). In dried produce, increases in the level of crude protein have been achieved but the relationship between protein and yield has appeared to be inversely proportional. Current work is engaged in identifying the genes underlying protein content and yield and improving this relationship (ProtYield, 2015).

The protein quality is in itself important as the presence of anti-nutritional factors can affect the growth rate of livestock. The development of low-vicine and convicine varieties of *Vicia* beans and low-tannin beans has been targeted specifically for crops destined for animal feed.

Resistance to pests and diseases

Disease resistance is another important goal for breeding new varieties, though many of the pathogens that are involved in root or foliar diseases have several strains, races or pathotypes. Many of the current commercial varieties have resulted through selection of breeding lines that exhibit a satisfactory level of field tolerance to disease, but in some situations crops become infected due to the presence of a mixture of pathogen races. This is especially the case with pea downy mildew, where populations of mixed races occur. Complete or partial resistance to diseases is a difficult objective. Root-infecting pathogens can also occur in mixed populations and resistance to one does not confer resistance to others, as in the case of the pea foot rot complex of *Fusarium solani, Aphanomyces euteiches* and *Didymella pinodella*.

Pesticide availability

The range of active ingredients used for pest, disease and weed control in legumes has become more limited as a result of a combination of factors, including the economics of production and development of pesticides for what is termed a minor crop, and environmental and health concerns over certain actives that have resulted in national or international restrictions on use. This has been especially felt in the European Union, where all available pesticides have been reassessed for safety issues and many have been withdrawn. The process is ongoing and the development of acceptable effective alternative actives has not kept pace with the losses. Similar programmes are in place in other countries. This is resulting in a very restricted number of pesticides being available for use in peas and beans (Andersons Centre, 2014). Alternative approaches to crop protection include mechanical weeding techniques, pest monitoring and avoidance, development of disease-resistant varieties and changes in the cropping rotations on-farm and regionally. It is, however, difficult to envisage large-scale commercial production of crops without the need for some pesticide products.

Drought tolerance

Although large-seeded legume crops are less demanding on water than cereals or other large-scale arable crops, drought has a large detrimental effect on growth and yield. Moisture-stressed plants are also more susceptible to disease and peas in particular are more likely to become infected by the soil-borne pathogenic fungus *Fusarium* f. sp. *pisi*, which causes various vascular wilt diseases. The basis for drought tolerance in peas is being studied using the related legume model plant *Medicago truncatula*, where the genome is now well known, to identify the genes responsible for drought tolerance. A small study in the UK examined the current varieties of combining peas and faba beans for their relative drought tolerance in a replicated trial under protection with and without irrigation. The study, which was carried out in 2009, clearly showed that there were significant differences between varieties under drought-stress conditions but less so in the case of faba beans (PGRO, 2009, unpublished). Work is also studying the relationship between drought tolerance and fungal infection in peas (ABSTRESS, 2013). Development of drought-tolerant varieties will further extend the range of soil types and cropping areas to enable reliable production.

Demand for animal feed

As stated earlier, the increasing demand for meat and dairy products in developing countries has significantly increased their demand for feed that is based largely on soybean protein. Faba bean and pea have been proven to provide a useful level of protein and energy which, although not as high as soya, nor is the amino acid profile completely comparable, nevertheless is being used widely by animal feed compounders; in addition, an increasing amount is being used for aquaculture, particularly salmonids, to replace soya and some of the fishmeal used by this industry. This is already having an effect on the areas of crop planted, particularly in Europe, and this situation is expected to continue for some years. Any increase in cropping will inevitably encourage breeders to continue to develop pea and bean varieties for temperate areas of the world.

Climate change

Evidence of a gradual increase of air temperature has been gathered by many sources throughout the world. Whereas some effects of climate change may result in a more beneficial environment for crop growth, the risk of extremes of weather within a season will doubtless have some deleterious effects. There is also evidence that migration of some pests and diseases is taking place to regions not normally at risk, putting greater pressure on crop production. Pea and bean growers in temperate areas are already noticing the effects with some varieties particularly susceptible to extreme weather events (NFU, 2010). Another aspect of the effects of increasing global temperature has been the acceleration of harvest date for spring-planted crops. A study by PGRO (S. Belcher and A. Biddle, unpublished) demonstrated that a single named variety of vining pea that had been sown in the early spring each year from 1983 until 2014 became ready for harvest approximately 3 weeks earlier over the period of time (Fig. 8.1). This could have had the effect of lowering the yield, as

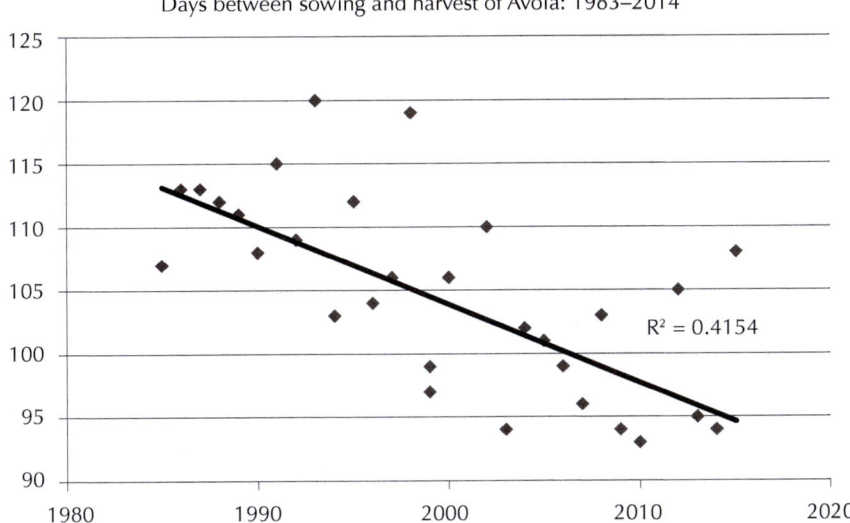

Days between sowing and harvest of Avola: 1983–2014

$R^2 = 0.4154$

Fig. 8.1. Trend of earlier harvest date of a single variety of vining peas (from S. Belcher and A. Biddle, unpublished).

the number of days between sowing and harvest was significantly reduced. However, there may be opportunities for growers in temperate areas to improve productivity by adopting earlier-maturing varieties that may be able to withstand higher summer temperatures and drier conditions. Soil temperatures may become more reliably higher in the spring, enabling earlier sowing regimes and earlier harvest, which may be more beneficial to vegetable production. There may also be opportunities to produce more novel crops such as dried *Phaseolus* beans in northern Europe. The drier areas, including Mediterranean countries, may suffer more damaging effects unless more drought-tolerant varieties become available.

Breeding technologies

The development of a new breeding line into a commercially viable variety takes many years and traditional breeding methods involve crossing and subsequent growing and selecting for a set of desirable traits; it can then take up to 12 years before commercial trials are put into place. The genetic mapping for *Pisum* has been all but completed but for *V. faba* this is at a relatively early stage. Pea genetics is rapidly evolving to utilize molecular-assisted approaches to find the molecular bases of those desirable traits in order to enhance breeding and shorten the time and number of generations needed to develop a finished variety (Smykal *et al.*, 2012).

A number of techniques are being used to identify key traits and these include the use of model legumes, particularly *M. truncatula* (Thompson *et al.*, 2009), as a base for identifying key genes controlling such traits as seed size and composition, drought and disease resistance (ABSTRESS, 2013) and others.

The use of fast neutron-induced mutants is also being studied (Domoney *et al.*, 2013) whereby a mutagenized population of peas has been generated to enable the identification and isolation of genes that underlie traits and processes.

Significant areas of pea and bean production continue to be grown, especially those varieties that are grown for dry harvest and are the principal source of protein in human nutrition. However, productivity has tended to plateau in these crops whereas productivity of other crops, including soybean and cereals, has continued to increase. Varieties of peas and beans with improved resistance to biotic and abiotic stress and with higher yield and protein content are required for these markets. Peas and beans have a narrow genetic base and although there is some scope through cross-pollination to extend this base, there is an argument that there is a need to use transgenic technologies to improve these crops.

Transgenics in peas and beans is at a relatively early stage of development but some significant advances have been made in certain aspects, such as resistance to pea seed bruchid (Shade *et al.*, 1994). Other characteristics such as glyphosate resistance and super-sweet types have been worked on. However, public concern over the use of transgenic production of food crops has resulted in a restriction of research and development and no commercial varieties have been released as yet.

As well as these techniques, there is still a large section of work using wild relatives of peas and beans to screen populations for certain desirable characteristics. This is particularly useful in picking out the disease-resistant relatives and identifying the underlying mechanisms involved (Warkentin *et al.*, 2015). Recently a new gene for resistance to powdery mildew has been identified from *Pisum fulvum* and introduced into peas; also varieties with resistance to broomrape have been released (Rubiales, 2014).

Plant architecture also has an effect on exacerbating or resisting the onset of diseases, particularly fungal pathogens such as *Didymella pinodes* (Baranger *et al.*, 2015). It is suggested that an upright standing stem with a reduced leaf area will create a microclimate within the crop that is unfavourable for fungal infection.

Legumes in the cropping rotation

The value of a crop that is able to synthesize its own nitrogen requirement and also leave a residue of nitrogen that is then available to a following non-leguminous crop has been well defined in earlier chapters. The cultivation of a crop that requires no additional nitrogen in the form of fertilizer also significantly reduces the amount of nitrous oxide that occurs when inorganic nitrogenous fertilizers degrade in the soil (Jeuffroy *et al.*, 2013).

The potential for developing non-legume crops that are able to utilize soil-borne *Rhizobium* by means of engineering the root system to produce nodules is an exciting concept that has already received significant research input (Rogers and Oldroyd, 2013).

Because the crops are able to produce their own nitrogen nutrition, peas and beans avoid environmental risks due to the manufacture, transport and use of chemical nitrogen (Munier-Jolain and Carroueé, 2005). The reduction in the requirement for nitrogen fertilizers also reduces the amount of nitrogen that is leached out of the soil during periods of heavy rain, as the nitrogen produced by peas and beans is assimilated efficiently because the symbiotic fixation is optimal during the growth of the crop.

In an arable cropping rotation that includes a legume crop, the total nitrous oxide emissions can be as much as 25% of the whole farm cropping sequence. There have been several schemes available in some countries, particularly those in the European Union, whereby the inclusion of a legume has been encouraged through financial incentives such as support prices for the produce and more recently through an environmental initiative to maintain a cropping rotation that encourages the use of large-seeded grain legumes (European Parliament, 2013). The rationale for such schemes accepts the value of legumes both as commercially viable crops and as a means of helping to improve the arable cropping environment. The immediate effect of such a regime in the EU was an increase in the area of peas planted since the regulations came into force in 2015 (Fig. 8.2.) but this pattern was not duplicated by all member states.

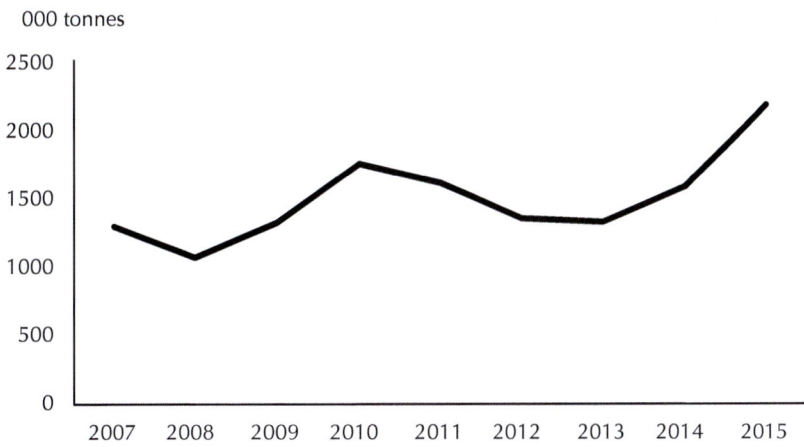

Fig. 8.2.   Total dry harvested legume production in the EU (source: Eurostat).

Food uses and processing developments

The demand for vegetable protein continues to increase in the developing world and already there is a diverse range of foods available for human consumption. Health aspects figure highly and peas and beans of most varieties and types are being utilized more than ever. The increasing number of gluten-intolerant consumers is also resulting in an increased demand for gluten-free alternatives in the baking industry and pea flour is an important ingredient in specialist breads and biscuits.

The freezing industry, particularly in Europe and the USA, seems to have reached a plateau, or in fact a slight decline in sales of frozen vegetables, including peas. However, the inclusion of peas in ready meals and other complete products such as pizza and vegetable mixes is more stable. Efforts are continuing in the promotion of frozen vegetables as a healthy part of the diet together with developments in packaging to reduce waste.

The rate of use of peas and beans for animal feed has already been mentioned but developments in processing technology are increasing the range of uses of the various components of peas and beans. Current studies are investigating the effectiveness of processes that can be used to separate the starch from the protein using de-hulled, milled faba bean flour (Beans4feeds, 2012). The flour is passed through a vertical cyclonic air stream, which separates the components of the flour by air classification based on specific weight. This provides proteins and starch-enriched concentrates, enabling components of different ratios of starch to protein for suitable diets for poultry and pigs and for ruminants. The work has also shown that bean protein is very useful in aquaculture (De Santis *et al.*, 2015). Such developments provide a more useful and less wasteful market for beans and would also be suitable for peas, providing further initiatives for pea and bean production.

THE FUTURE

There are many research programmes in progress that are aimed at overcoming the main constraints to pea and bean production both in the developed and in the developing countries of the world. All are aimed primarily at extending the yield and increasing the quality of the harvested product through plant breeding and agronomic management and at extending the demand through innovation for food for the growing human population, increasing human health and reducing the dependence on non-sustainable commodities, as well as food for animal and fish production. In addition, work is ongoing to exploit the crops as a source of pharmaceutical and nutraceutical products.

The world population is projected to reach 11 billion by 2100. Peas and beans have been an important part of the human diet for thousands of years but despite the fact that they provide a healthy and sustainable source of protein there has been only a slight increase in production worldwide, unlike other major crops such as cereals, rice, maize and soya, which have had increases of production of 200–800% over the past 50 years. Future crops must be developed to provide varieties that are adaptable to a wide range of growing conditions, can withstand periods of high or low soil moisture availability, must be resistant to commonly found pests and diseases and reduce the risk of variability of production from year to year.

Consumption has been declining slowly in both developing and developed countries as demand for meat and dairy products grows. However, such an increase in animal and dairy products as well as a rising demand in farmed fish is putting a strain on the more traditional sources of plant protein, namely soya. The role of pulse as a soya substitute is beginning to increase and recent work in the developed nations has shown the value of peas and beans as soya replacement in rations for pigs, poultry, fish and ruminants.

This demand is already encouraging a growth in international trade. Large producers such as Canada are supplying product to India and the Far East at increasing levels. This trend is continuing and a country such as China, which was self-sufficient in dry harvested legumes, is now a net importer as the price of soya beans and soya meal continues to rise (FAO, 2016).

The future for peas and beans as adaptable, sustainable, environmentally friendly and with a high production potential, together with an increasing demand, is positive.

REFERENCES

ABSTRESS (2013) *Improving the Resistance of Legume Crops to Combined Abiotic and Biotic Stress.* EU Seventh Framework Programme Project number FP7-KBBE-2011-5-289562. Food and Environment Research Agency (Fera), York, UK.

Andersons Centre (2014) *Crop Production Technology: The Effect of the Loss of Plant Protection Products on UK Agriculture and Horticulture and the Wider Economy.* Report prepared for Agricultural Industries Confederation (AIC), Crop Protection Association (CPA) and National Farmers Union (NFU) by The Andersons Centre, Melton Mowbray, UK. Available at: www.nfuonline/assets/37178 (accessed 13 February 2017).

Baranger, A., Giorgetti, C., Jumel, S., Langrume, C., Moussart, A. *et al.* (2015) Plant architecture and development to control aerial disease epidemics in legumes: the case of Ascochyta blight in pea. *Legume Perspectives* 7, 19.

Beans4feeds (2012) Beans4feeds represents a £2.6m research investment and is an 11 partner industry led and co-funded research project with the UK's Innovate UK. Available at: http://beans4feeds.hutton.ac.uk (accessed 13 February 2017).

De Santis, C., Ruohonen, K., Tocher, E., Secombes, C.J., Bell, J.G. *et al.* (2015) Atlantic salmon (*Salmo salar*) parr as a model to predict the optimum inclusion of air classified faba bean protein concentrate in feeds for seawater salmon. *Aquaculture* 444, 70–78.

Domoney, C. (2011) Understanding pea seed quality. *The Vegetable Magazine*, Winter 2011, p.5. Processors and Growers Research Organisation, Peterborough, UK.

Domoney, C., Knox, M., Moreau, C., Ambrose, M., Palmer, S. *et al.* (2013) Exploiting a fast neutron mutant genetic resource in *Pisum sativum* (pea) for functional genomics. *Functional Plant Biology* 40(12), 1261.

European Parliament (2013) *The Environmental Role of Protein Crops in the New Common Agricultural Policy*. Directorate General for Internal Policies, European Parliament. Available at: http://www.europarl.europa.eu/RegData/etudes/etudes/join/2013/495856/IPOL-AGRI_ET(2013)495856_EN.pdf (accessed 13 February 2017).

FAO (2016) *International Year of Pulses – Nutritious Seeds for a Sustainable Future*. Available at: www.fao.org/pulses-2016/ (accessed 13 February 2017).

Jeuffroy, M.H., Baranger, E., Carrouée, B., de Chezelles, E., Gosme, M. *et al.* (2013) Nitrous oxide emissions from crop rotations including wheat, oilseed rape and dry peas. *Biogeosciences* 10, 1787–1797.

Munier-Jolain, N. and Carrouée, B. (2005) Prospects for legume crops in France and Europe. In: Munier-Jolain, N., Biarnes, V., Chaillet, I., Lecouer, J. and Jeuffroy, M.H. (eds) *Physiology of the Pea Crop*. INRA-Arvalis Institut du vegetal UNIP ENSAM, Paris, pp. 239–242.

NFU (2010) *Focus on Peas and Beans: Climate Change Series*. Fact sheet 13, Farming Futures. Available at: www.foodandfarmingfutures.co.uk (accessed 13 February 2017).

NIAB TAG (2014) *Landmark Bulletin*, January 2014. National Institute of Agricultural Botany, Cambridge, UK.

ProtYield (2015) Protein content vs yield in legumes: releasing the constraint. Project co-funded by the Technology Strategy Board, in partnership with Defra and BBSRC, UK. Pulse Crop Genetic Improvement Network (PCGIN), Norwich. Available at: http://www.pcgin.org/ProtYield/ProtYield.pdf (accessed 13 February 2017).

Pulse Canada (2008) *Pulses and Diabetes Control*. Fact Sheet. Pulse Canada, Winnipeg, Canada. Available at: http://www.pulsecanada.com/uploads/cc/4d/cc4d808b3ed8584bbda594e400274d38/11-Oct-5-Diabetes-fact-Sheet-FINAL.pdf (accessed 13 February 2017).

Rogers, C. and Oldroyd, G.E.D. (2013) Synthetic approaches to engineering the nitrogen symbiosis in cereals. *Journal of Experimental Botany* 65(8), 1939–1946.

Rubiales, D. (2014) Legume breeding for broomrape resistance. *Czechoslovakian Journal of Genetic Plant Breeding* 50(2), 144–150.

Shade, R.E., Schroeder, H.E., Pueyo, J.J., Tabe, L.M., Murdock, L. *et al.* (1994) Transgenic pea seeds expressing the α-amylase inhibitor of the common bean are resistant to Bruchid beetles. *Nature Biotechnology* 12, 793–796.

Smykal, P., Coyne, C., Ellis, N.T.H., Flavell, A.J., Ford, R. *et al.* (2012) Pea (*Pisum sativum*) in the genomic era. *Agronomy* 2(2), 74–115.

Thompson, R., Aubert, G., Duc, G., Gallardo, K. and Lesignor, C. (2009) *Medicago truncatula* as a model legume. *Grain Legumes* 53, 5.

Warkentin, T., Smykal, P., Coyne, C.J., Weeden, N., Domoney, C. *et al.* (2015) Pea (*Pisum sativum* L.). In: De Ron, A. (ed.) *Handbook of Plant Breeding: Grain Legumes*. Springer Science and Business Media, New York.

INDEX

Note: Page numbers in **bold** type refer to **figures**
Page numbers in *italic* type refer to *tables*